—TRUST—

SELF-INTEREST AND THE COMMON GOOD

Marek Kohn is an Honorary Faculty Fellow in the Centre for Applied Philosophy, Politics and Ethics at the University of Brighton. He lives in Brighton, England, with his wife and son, and has published five previous books: *A Reason for Everything: Natural Selection and the English Imagination* (2004); *As We Know It: Coming to Terms with an Evolved Mind* (1999); *The Race Gallery: The Return of Racial Science* (1995); *Dope Girls: The Birth of the British Drug Underground* (1992/2003); and *Narcomania: On Heroin* (1987).

Trust

SELF-INTEREST AND THE COMMON GOOD

Marek Kohn

OXFORD
UNIVERSITY PRESS

OXFORD
UNIVERSITY PRESS

Great Clarendon Street, Oxford OX2 6DP

Oxford University Press is a department of the University of Oxford.
It furthers the University's objective of excellence in research, scholarship,
and education by publishing worldwide in

Oxford New York

Auckland Cape Town Dar es Salaam Hong Kong Karachi
Kuala Lumpur Madrid Melbourne Mexico City Nairobi
New Delhi Shanghai Taipei Toronto

With offices in

Argentina Austria Brazil Chile Czech Republic France Greece
Guatemala Hungary Italy Japan Poland Portugal Singapore
South Korea Switzerland Thailand Turkey Ukraine Vietnam

Oxford is a registered trademark of Oxford University Press
in the UK and in certain other countries

Published in the United States
by Oxford University Press Inc., New York

First published 2008
First published in paperback 2009

British Library Cataloguing in Publication Data

Data available

Library of Congress Cataloging in Publication Data

Data available

ISBN 978–0–19–921791–5 (Hbk.)
ISBN 978–0–19–921792–2 (Pbk.)

10 9 8 7 6 5 4 3 2 1

Typeset by SPI Publisher Services, Pondicherry, India
Printed in Great Britain
on acid-free paper by
Clays Ltd, St Ives plc

CONTENTS

PREFACE

Now that agreement has been reached about how humankind can best make a profitable living, with a single economic orthodoxy established around the world, an increasing number of scholars and commentators have turned their attention to the question of how people can live well. Recognizing that beyond a certain level of material security, money delivers diminishing returns, they reflect on what really makes life rewarding, and what makes a society good, rather than merely prosperous. They write about qualities like community, status, respect, and happiness. Trust is one of these qualities. Like its companions, it is fundamental to a fulfilling life and a good society.

This short book is intended as a contribution to the growing discussion about these precious, and elusive, qualities of life. It is not a survey of the literature in the academic sense, nor a pocket textbook, nor even a primer, but an essay that responds to a rich array of knowledge and ideas. I have been intrigued, challenged, and inspired by what I have read in researching it: I have written it in a way that I hope will evoke similar responses from its readers.

It starts with individuals and moves on to nations. The terrain is sketched out in Chapter 1 with an everyday scene that illustrates the main personal and public dimensions of trust. This opens up an introductory discussion about what trust is, what it feels like, and what conditions are required for it to develop.

The argument then re-starts in Chapter 2 from first principles, asking how cooperation can evolve among organisms or other agents pursuing their individual interests. Discussing some of the answers to this question that have been obtained by the application of game theory, it describes a remarkable example of real-life games played by soldiers across enemy lines in the First World War. Chapter 3 also begins with evolutionary theory, considering the problem of how signals produced by self-interested organisms can be reliable, and ends with a philosophical discussion of whether trust can be rational.

In Chapter 4 the relationship between authority and trust is examined, describing the course taken from traditional authority to modernity. Starting with a discussion of trust in gods, the account goes on to examine how people have sought to invest their trust as the power of religious faith and associated traditions has diminished. Although trust is extensive in the experts who create the systems through which much of modern life is lived, it is far from perfect, and is compromised by undercurrents of distrust in institutions.

Chapter 5 then considers trust in society, or its absence, the effect of low social trust on economic development, and the idea of social capital. It reflects on the polarization of trust between mutually antagonistic communities, each warming itself by the flames of its hostility to the other, and on trust in districts shared by members of diverse groups. The relationship between social trust and people's trust in political institutions is discussed in Chapter 6. Communist totalitarianism relied on distrust, and devoted much of its energy to orchestrating it among its subjects. Liberal democracy is also based on distrust, but in the opposite direction: it is founded upon the suspicion that the powerful will be tempted to abuse their power, and so must be subject to checks and balances. Among its citizens distrust is endemic, but only to the point of complacency.

This theme is concluded in Chapter 7, which points out some of the weaknesses of the idea of generalized trust, while affirming that it expresses a quality that is fundamentally important to a good society. The chapter and the book then end with reflections on the place of trust in a world that is in constant flux, spinning ever faster, compulsively initiating, revising, rearranging, and discarding its relationships.

I am most grateful to Latha Menon for commissioning and editing this book; to Charles Lauder Jr, Eva Nyika, and James Thompson for helping to realize it; to Andrew Brown, Christine DeBlase, Gavin Keulks, Natalia Madjarevic, David Skinner, and John Street for advice and assistance of various kinds; and to three anonymous readers for their very helpful comments on the original proposal.

1

JUST GOING ROUND TO THE SHOP

THE KITCHEN is in full steam; pans hiss and spoons clatter, plates are slapped onto trays and sauces marshalled; the parent animating all these objects leans across and opens the breadbin—which, being empty, stops the cavalcade in its tracks. She steps out to reach for her coat, then pauses again; instead, she calls to summon her child, causing a hiatus in another sphere of activity, whose clattering, marshalling, and animating constitute the equally busy domain of play. She hands him some coins and sends him out, down the street and across the road to the shops. Within a few minutes he returns, carrying a loaf; behind in the shop he has left the payment and a recorded video image of himself. In this brief and everyday episode, the two of them have negotiated or touched upon most of the basic dimensions of trust.

Two transactions have taken place: the one between parent and child, the second between child and merchant. The second belongs to the class that has preoccupied many theorists of trust, those whose vision is set in the marketplace, and who conceive of trust primarily as an issue that arises between individuals idealized as rational, free, and equal. Starting here, within the home and the family, we get our priorities in order.

Trust begins, or should begin, between parent and child; it is upon the basis of trust thus established that individuals can go out to become actors in a marketplace. And starting here, with a child offering a coin to a shopkeeper, we set off aware that for most of the transactions of everyday life, trust must inhere in relationships between agents who are significantly unequal in power, resources, or autonomy.[1]

Before the child sets off out of the door, there must be trust between him and his mother. On his part, this is based on the unconditional and total trust that arises—unless suppressed by neglect or gross distortion of parental roles—as a founding condition of childhood. He is old enough to be able to disagree with a parent about the details of where his interests lie, but his sense of his place in the world rests upon his absolute confidence that his parents have his fundamental interests at heart. He is sure that his mother would not send him into a world in which he would be in peril. More than either of them are consciously aware, his sense of basic security is created and sustained by the routines of life: the Saturday breakfast, more relaxed than on weekdays, with its menu compiled by the family's various members, the iterated children's games that put a familiar cast of characters—plush, plastic, or pixels—through variations upon a basic repertoire of parades and adventures. All these affirm that the world may be treated as if it were constant, reliable and secure.

That goes for his mother too. Her everyday decisions must be based on a sense of normality; that the world is constant enough for its risks to be intuitively assessable. She must feel a basic safeness underneath the inevitable risks; she must feel some degree of basic trust in people whom she does not know personally or at all.

She must also trust her child, in two successive respects. First she must be sure of his competence to do what she asks of him. He must be old enough to be capable of the task. If so, he is also old enough to be left in the house while she runs the errand herself. Indeed that's her initial impulse, and would probably be simpler, but she has strategic considerations to weigh. He needs to learn how to go out and act in the world; she must learn to trust him to do so. She must enable him to become trustworthy.

Once she trusts in his competence, she must trust him to apply that competence in her interests. She trusts him not only to collect the change but to return it to her in full; she trusts him to buy what she has specified rather than what he might prefer. Her faith in him may be augmented by norms that he has absorbed, from school or other public sources as well as his parents, encouraging him to pick up a brown loaf instead of a sugar-laced white one or a carton of doughnuts. His choice is underwritten by his keenness to do what she wants him to do; her interests are thus incorporated in his.

She must also trust him not to trust. Her watchword will be 'Take care crossing the road, and don't talk to strangers.' Distrust of strangers is regarded as a precondition of independent ventures into public space; and even road safety involves awareness that relates to trust. At the crossing the child must remember that cars may not always stop when the lights change. When he looks up at the security camera overlooking the till—or the street—he will see a reminder that public spaces contain untrustworthy people. And these include children. The shopkeeper is prepared to trust him on his own, but would not if he came in with a few of his friends—'No more than four schoolchildren at a time' is the quota stipulated on the door.

Nevertheless, these anxieties rest upon a thick carpet of trust. All concerned trust that the traffic lights will not suddenly malfunction and send vehicles off into collisions. The shopkeeper and the customers trust that the till calculator will reckon the bills correctly and that the banks will record the resulting changes in balances accurately. They trust the money; which is to say that they trust the immense and mysterious network of systems and institutions that guarantee the banknotes' promises to pay their bearers what they say they are worth. This implies that whatever complaints they may have about the government, they have a basic confidence in the functioning of the state and the economy. Impersonality seems to promote trust in many contemporary contexts. The customers will ignore the automatic camera, but if they saw the shopkeeper photograph the child himself, they would probably call the police.

We can recognize this as a distinctively contemporary cameo not just because of the presence of a security camera, but because it is infused with a sensitivity to risk that would seem neurotic to earlier generations. Contemporary analyses of trust emphasize its intimate connection with risk, and some go so far as to depict trust as a specifically modern phenomenon arising from a preoccupation with risk that has arisen as this particular phase of history has developed. Trust does not arise when tradition dictates each person's place and how they shall conduct themselves in it. When your actions are determined by your station in society, your gender, your parents, their parents, and their parents before them, all implementing a universally accepted, pitilessly enforced body of rules that prescribe your actions in any situation you are likely to encounter, then what's to trust?

The answer is that life is never so ordered or choices so forced that the need for trustworthiness can be eliminated. A society may be segmented by rank and bound by codes of honour, yet allow Machiavellian individuals to flourish—not least Niccolò Machiavelli himself, in Renaissance Florence. In any society there will always be scope among peers, such as friends, siblings, or trading partners for choices about the extent to which one takes another's interests into consideration. That would have been the case when people all gathered or hunted their food, rather than growing or buying it. Questions of trustworthiness arose as humans became human, if not before.

That is not to deny a distinctive character to modern relationships, and the questions of trust that arise in them, however. They differ from traditional ones in number and duration. Throughout human prehistory, most of history, and large areas of today's world, people have generally tended to interact with small numbers of other people over extended periods of time. For the most part they would spend their time with people of their own small group, and their contact with outsiders would be limited. In the modern world the reverse is becomingly increasingly true. People range further, in person or via electronic networks, and frequently have encounters with people they never meet again. Their circles of acquaintance are continually being updated.

At the same time, the numbers in the inner family circle have fallen. Perhaps the child sent to the shop has siblings; perhaps they're out with their father; but there'll only be a few of them compared with the traditional chorus. This, rather than bureaucratic angst, is what sets the terms for contemporary assessments of risk. Large parts of the world are now reliable enough to permit people to have very few children, and to concentrate on them enough to enable them to take advantage of the opportunities that the world now presents. People no longer accept that they should have many children and expect to bury one or two of them; they need to believe that the few children they have will grow up safely to live lives that should be long, healthy, comfortable, and rich in experience. The risks both of loss in an absolute sense and of lost opportunities are unconscionable, and periodically fray the fabric of public trust.

Often what picks it apart is food. Buying a loaf requires implicit trust in the systems that ensure the safety of food. It need not be absolute trust: a parent who prefers a brown loaf to a white one may also be bothered about the addition of various chemicals to the bread. But it is sufficient... until the food in question becomes the focus of a health scare, at which point ambient trust may be replaced by incandescent suspicion. Blame will then find its way to the government, as inevitably as water flows downhill, highlighting the mistrust of political institutions that helps to define the modern political condition. Much of the attention devoted to trust in public discourse revolves around the public's suspension of belief in political theatre. This may, however, be a problem for politicians rather than for democracy, whose structural foundations are actually based on mistrust. Liberal states take as given that those with power will be tempted to abuse it, so the powers of the state are separated into divisions intended to check and balance their exercise.

Negotiations between presidents, premiers, parliaments, and courts are relatively clear, from the public's point of view, if remote. Alarms about health and safety are more disturbing, because they raise questions not only about governments and bureaucrats but about the expert systems on which modern life depends. Wrangles in the upper strata of power, for instance between a government and the judiciary, appear usually to be over values rather than facts. Health scares throw facts into question. They raise doubts about networks that are even less comprehensible than the money system, involving industrial processes and scientific investigations into the effects of substances or pathogens. The public would like to see these systems as disinterested and reliable, like bank transactions or air traffic control, but is uneasily aware that experts differ, science proceeds by debate, and commercial and political interests are inevitably at work. When public alarms arise in specifically medical contexts, where people feel that their lives are in professionals' hands, mistrust may rapidly flare into a sense of betrayal.

The child has come back from the shop with a loaf and a trail of actual and potential issues of trust. They begin with the original trust between parent and child; they resolve into questions of competence and commitment; they include a background of trust in expert systems; they touch upon anxieties about malevolence and dishonesty in public spaces; they point to the volatility of public trust in certain institutions. The scene works as a three-hander—parent, child, trader—but in doing so poses questions about the rest of the cast. The other parent could be round the corner, or could be anywhere. Others with varying degrees of significance could be part of the family network. Families and intimate relationships are short of certainties and full of possibilities these days, their variety of forms creating a new variety of demands for trust.

Making sense of trust requires deciding what trust is. Many theorists have devoted themselves to this task over the past two or three decades, and have arrived at many different conclusions. Instead of trying to resolve their differences, it may be helpful to reflect briefly upon the community of meanings labelled 'trust'; to inquire whether its various senses are close kin, tightly bound factions or passing acquaintances. At this stage description and definition may be equally useful.

We might begin with an attempt to specify the formal terms of a trusting relationship: A trusts B to do X. Envisaging trust as a relation between two particular parties, concerning a particular action or range of actions, this formulation (by Russell Hardin) will define trust for the purposes of this book.[2] But instead we'll start with what trust feels like. Even the word has the gift of warming the heart and dissolving its tensions. Very few other words have that power. 'Grace' is one, but remains available only to those who have retained the necessary religious faith. The other is 'love', which is often related to trust, and akin to it in quality.

Like love, trust is involuntary. You can't help whether you love or trust someone; you either do or you don't. Although situations may arise where one says, 'I'll just have to trust you then', that is not really trust. Rather it means 'I'll just have to take the chance that you will act in my interests,' or 'I will have to act in the same way that I would if I really trusted you.'

These reserved declarations, however, may be steps on the way to trust. Sometimes one says, 'I'll just have to trust you' ironically: it is an opportunity to acknowledge that A already trusts B, though it may not have been explicit up to that point. In other circumstances 'I'll just have to trust you' is an invitation to prove yourself trustworthy. These subtexts are initiatives undertaken in the hope of establishing or deepening trust. Although you can't make yourself trust, you can act in ways that help trust develop; as is also the case with love.

Involuntary feelings are often passionate ones, and that is true for certain celebrated forms of trust, such as those between lovers or comrades. Trust may be unconditional too—but rarely so outside the special relationships between parents and young children. Normally trust is conditional and limited. It is a practical attitude rather than a transcendent passion. At the same time it is also, of course, optimistic. Trust is an expectation, or a disposition to expect, that another party will act in one's interests.

It could be termed confidence, or reliance, as in the *Oxford English Dictionary*'s first definition of trust: 'confidence in or reliance on some quality or attribute of a person or thing, or the truth of a statement'. Some theorists distinguish between trust and confidence, reserving trust for agents with intentions and confidence for things or processes. 'I do not trust the sun to rise each day, at least not in any meaningful sense beyond merely having great confidence that it will do so,' observes Russell Hardin; who is also careful to distinguish between trust and trustworthiness, the latter being what is indicated by qualities and attributes.[3] But although some might find it wanting in precision, the *OED*'s definition is not at odds with the view that trust is an expectation about the actions of others. Its last part serves as a reminder that of all the actions that trust is concerned with, few are more critical than the act of telling the truth.

When we expect others to act in our interests, we do not expect them to act against their own interests. As Russell Hardin proposes in his 'encapsulated interest' model of trust, we must believe that their interests incorporate ours.[4] When a parent sends a child on an errand to buy a loaf, the coincidence of interest is all but total. Both are concerned before all else with the safety of the child; both want to obtain the loaf, which they will both consume. The potential conflicts of interest are marginal and matters of interpretation: the child might prefer not to bother, or might prefer the tastier white bread over the healthier brown, but his mother will readily make the case for why doing as she asks is good for his health and character. Negotiations such as these will help to embed her preferences within his.

We can be sure about trust and the intertwining of interests in this case because we can be in no doubt that there is a relationship between the parent and the child. There may well be relationships between each of them and the shopkeeper too: he is likely to know his local customers at least by sight; he may enjoy his day-to-day encounters with them; and he will certainly consider that it is in his commercial interests to keep their custom. If the relationships between trader and customers remain healthy, the latter's interests will be at the heart of the former's.

Out in the street, however, interactions are usually too brief to be described as relationships. Road sense must be based around a precautionary principle of mistrust. The child must learn that signals and actions do not always correspond: occasionally a car will go through a red light, and cars turn without signalling all the time. But people frequently seem to cite driving as a striking example of trust in strangers. We are impressed at our readiness to take our chances among columns of massive steel projectiles, each controlled by a mind of its own. It is true that driver A's interests incorporate driver B's to the extent that each wants to avoid colliding with the other. They do not normally know each other, however; they may not communicate with each other, and in most cases are only peripherally aware of each other's presence. If one considers that trust involves particular expectations about particular agents, drivers don't trust each other. Their trust is placed in the technical systems that control the flow of traffic; their confidence is built by the range of punishments that traffic systems threaten against drivers who break the rules; and they usually assess the risks of driving as acceptably remote. In some circumstances some drivers may be deterred from making journeys because they consider that either roads or drivers are likely to be more dangerous than usual: in fog or on Saturday nights, for example. But these are essentially perceptions about how other drivers balance their own interests, between the urge to arrive quickly and the risk of not arriving at all, when their judgement is tested by poor visibility or impaired by alcohol. Under those conditions driver A is even less aware of driver B than usual.

Although trust may require a meaningful relationship to satisfy its more demanding analysts, it need not require goodwill. When A is a person and B is a bank, A may trust B to keep her money safe although she does not imagine for a moment that the bank feels warmly disposed to her, and she may well suspect that it will assert its interests at her expense when it gets a chance to levy charges or manipulate interest rates. If A enters hospital and is examined by Doctor B, she may trust B's professional expertise and integrity even though B appears indifferent to her as a person. Until relatively recently indifference on the part of medical professionals, or plain rudeness, was if anything regarded as a sign of trustworthiness: it implied the objectivity needed for expertise, and asserted the superior status that medical expertise conferred.

Trust can even be achieved between parties who are at war with each other. On the Western Front in the First World War, where armies were stopped up in immobile trenched lines for years, units facing each other negotiated covert truces. These were covert because they had to be concealed from the high commands on each side, and they were often negotiated by gunfire. Artillery shells would be fired into empty ground or at regular times; riflemen would shoot high or wide. Observing such signals a soldier could come to believe that the man facing him across the lines 'ain't a bad feller'.[5] Soldiers came to trust their opposite numbers not to attack them unless forced to do so by senior commanders. These informal arrangements (of which more later) were possible when units faced each other long enough for relationships to develop, even though these were relationships based on exchanges of fire.

Elsewhere in the Great War, trust in enemy soldiers was officially organized. One former prisoner of war later recalled how his German captors permitted officers to leave their camps on 'honour walks', having signed forms stating that they would not try to escape. On their return the officers would hand in the forms—and would be released from their undertaking.[6] Between these two examples we can see the old order passing and a new modern order being born. The First World War dealt traditional codes of military honour a blow from which they never recovered: the imprisoned officers whose word was sufficient to underwrite their privileges were among the last to enjoy such benefits. In the trenches, the rank and file had to work out for themselves how to create trust under modern conditions. They had marched to war in a state of patriotic enthusiasm that approached ecstasy; and found themselves in an unprecedented situation that differed radically from what they had been told they were entering. They set aside jingoism, propaganda, and unquestioning obedience in order to respond to conditions as they found them. The modern world made a mockery of traditional authority.

Presenting new and unpredictable situations at every turn, it demanded pragmatism and negotiation case by case.

Already it is clear that we can speak of trust in a great variety of circumstances, and with varying degrees of conviction. Trust that is part of intimate and loving relationships—parents and children, partners who are also good companions, confidants, close friends—feels like the real thing. So in its very different way does the kind of trust that takes impersonal systems—banks or airline operations—for granted. It is related to a more diffuse confidence in the trustworthiness of people and other agents that are encountered in the course of normal life, such as shop assistants or passers-by from whom one asks directions. At a further remove stand the kind of expressions of trust, or distrust, elicited by opinion surveys probing people's attitudes to organizations or their personnel, such as the police, doctors, or politicians. Then there are the colloquial usages, in which one might speak of trusting other drivers or 'having to trust' someone, that fall short of a firm purchase on trust's core meanings.

Those core meanings themselves become harder to grasp the more one tries to specify what degree of freedom makes trust possible. It is easy enough to agree that the word 'trust' is applicable in situations where the trusted party has some choice over whether to behave as the truster expects, and that it lacks conviction in the absence of such choice. A slave-owner in the Americas might have been completely confident that his slaves would act in his interests, but that confidence would be based on control. For many authors, the meaning of trust centres upon the willingness to accept vulnerability, the risk of suffering injury or loss if the trusted parties do not act as they are trusted to do; whereas slave-owners had chains and weapons to ensure their invulnerability.[7] Men living in certain traditional cultures today may be confident that their daughters will not form relationships with men except when they marry, which will be as their families decide and their cultures' rules dictate. Women who refuse to comply may be subjected to draconian punishment, including murder. That these punishments are an ordered element of traditional discipline, rather than uncontrolled explosions of anger, demonstrates that the system is not based on trust.

They also show that choice is possible, even though it may incur the ultimate price. Intuitively, such situations seem utterly different from ones in which people are free to choose their partners without fear of murder, but on the scale of constraints it is hard to draw a line separating situations where choice is possible and ones where it is not. Although European slave-owners may have needed chains in the Americas to control people abducted from Africa, that was because they had not sufficiently colonized the minds of their captives. The Catholic schools in Ireland that became notorious for their brutal regimes exercised day-to-day control by means of beatings, but they achieved far greater effects by the inculcation of guilt, which could last a lifetime.

The beatings helped to inculcate the guilt, though, and likewise it is impossible to separate the effects of physical violence and mental pressure, coercion and the willing internalization of norms. The constraints imposed by rules, obligations, customs, living kin, and dead ancestors raise not just the question of what to trust, but of whether trust really exists. People living in traditional cultures are likely to have a set of constraints on their choices of action that seem stifling to those accustomed to a culture that makes its rules up as it goes along. It is not immediately clear, however, that a person in one group is in a qualitatively different position from somebody in the other. The constraints in the traditional culture may be more limiting and the sanctions for transgression harsher, but the way norms work in any culture are similar. People are inhibited from certain courses of action, and induced towards others, by their beliefs about what is right, wrong, proper, and improper, by their desire to be good according to their culture's lights, their desire for respect and acceptance, and their abhorrence of shame.

Similar issues arise in the closest of relationships. We may say that a parent may be trusted never to harm her child, because we believe that she loves her child as parents should, that she is competent to raise a child, and that her behaviour will remain consistent because she is mentally stable. Her child is dearer to her than anything; she could not bear the child to come to harm, or the thought that her actions could put their relationship at risk; she believes that her duties to her child are her greatest responsibilities; she believes that the neglect or abuse of children are among the most abhorrent of wrongs; and she can imagine the shame and the ostracism that they would bring down upon the offender. So does she really have a choice in the matter? If not, how meaningful is it to describe her as trustworthy?

We could continue this line of questioning until it became clear that the question of whether trust really exists is a local instance of the question of whether there is such a thing as free will. The idea of trust depends on the assumption that agents really do have choices. Trust is an expectation about another's actions, based on the understanding that the other has the capacity to create mental models of possible courses of action, and to evaluate them within a framework that can incorporate interests besides the other's own. It is accompanied by the perception that the other's conclusions are not foregone, that genuine choices are being made. We cannot help but think this way. Even if lengthy philosophical analysis persuades some of us that free will may be an illusion, it still feels real. The perception of agency is fundamental to human understanding of the world. Within the realm of human relations, trust is one of its most significant effects.

Trust varies greatly in quality and degree. Much of the effort to analyse trust over the past several decades has been directed towards understanding which circumstances will promote trust and which will stifle or twist it. Plenty of significant factors will become apparent as the child walks round to the shop, starting with the fact that he lives close enough for it to be a neighbourhood shop. The more convenient its location, the more often he, his family and his neighbours will visit it; increasing the frequency with which neighbours meet each other and are encouraged to feel themselves to be a community. As regular customers they will become part of the shopkeeper's community, which should—all things being equal—encourage him to trust them.

The less equal things are, however, the worse become the prospects for trust. Cities in wealthy nations contain areas of relative poverty in which there are significant everyday risks from individuals or groups who pursue their interests by threat, theft, or force. Gangs intimidate their way, often simply by their presence, into monopolies of public space. Individuals or groups steal from shops or in the street, either covertly or through coercion. The threat of unprovoked violence may be increased by the presence of intoxicated or mentally disturbed people, in a setting that does not make it easy to moderate their behaviour. In such neighbourhoods it is unwise to assume that people are trustworthy. Many of them surely are, to the extent required for the quiet enjoyment of everyday life, but relations between them are inhibited by the pressure to avoid encounters with the untrustworthy ones. People are placed permanently on the defensive, particularly when, as in the case of shopkeepers, their daily lives consist of encounters with large numbers of individuals who are largely unknown quantities.

Uncertainty tends to be countered by assumptions. People make predictions about the character, disposition, and likely behaviour of others by regarding them as members of groups with defining and predictable characteristics. They may consider some groups to be threatening, some to be genial, some to be lazy, some to be industrious. Shops may become the focus of ethnic tension, especially when they are identified with a particular ethnic group. Resentment against such minorities, typically involving accusations of exploitation or disrespect, may sometimes erupt into riots. One section of the population finds grounds to agree, to cooperate, and to bond, by turning against another part of what might otherwise become a community.

This process, of developing solidarity within a group by identifying other groups as enemies, seems to be fundamental to human societies. Rooted in a sense of kinship, actual or imagined, it may create a kind of secure perimeter within which mutual interests are asserted and trust may develop. People who feel they share values and customs will tend to feel that they can predict each other's behaviour, and will be correspondingly confident that they can judge when others are trustworthy. If they feel that they have interests in common, they should be inclined to trust each other on matters arising from these mutual interests.

A sense of shared attitudes and interests, contrasting with those held by outsiders, may be promoted by stereotypical perceptions, popular prejudice, or propaganda. As unpopular governments often appreciate, one of the most effective ways for a faction within a group to obtain support from unsympathetic or antagonistic members is to engineer conflict with another group. By ensuring that there is a genuine threat from outside, the faction creates genuine and dominating common interests for all members of the group. The graver the threat, the greater the common interests: the potential for trust is correspondingly increased, though in practice it may only partly offset the mistrust that may arise over conflicts for scarce resources or suspicions that some are shirking their duties to the group.

Where antagonism results in physical separation between neighbouring groups, trust may flourish and doors be left unlocked within each perimeter. The inhabitants of each neighbourhood are not safe to cross the line, even though they may be all but indistinguishable to outsiders and even to each other. Aggressive sectarians in Northern Ireland have little to go on if they are seeking victims outside established Catholic or Protestant areas: names, addresses, and football allegiances are all cues that can be concealed or faked. The scarcity of reliable signals makes questions of trust particularly fraught for taxi drivers, who face additional hazards on top of the risks inherent to their occupation. They must make snap decisions about the trustworthiness of their fares, who may be aggressively sectarian, dishonest, members of illegal paramilitary organizations, or indeed all three. As an absorbing study by Diego Gambetta and Heather Hamill illustrates, Belfast taxi drivers must learn as much about trustworthiness as they do about street maps. Yet one of the researchers' most striking findings is that Belfast taxi drivers are more trusting and confident than their counterparts in New York. Belfast drivers have a sense of place and belonging. The province's ingrained patterns and adamant rules are the foundation of both their hazards and their security. New York taxi drivers, by contrast, are often recent immigrants with meagre community roots. The threats they face are hard to assess, lacking the rhyme or the reason of sectarian strife.[8]

Despite its apparently narrow basis, Northern Ireland's communal division has proved formidably sustainable. Although the main armed groups have discontinued their campaigns, and institutional biases against Catholics have long since been redressed, the social separation between working-class Catholics and Protestants has if anything grown deeper. At the political level, the Democratic Unionist Party has approached the 'peace process' with the utmost suspicion, and has become the leading party in the province. DUP leaders scorned their Ulster Unionist Party rivals' preparedness to take the secessionist Republican movement 'on trust', demanding proof that the Irish Republican Army had 'decommissioned' its arsenal as agreed.[9] They made it abundantly clear that although they might be prepared to cooperate with the Republican politicians of Sinn Féin, they could not imagine trusting them enough to take their eyes off them for a second. This radical mistrust is one of the reasons why nine years elapsed between the Good Friday Agreement, which established the terms of a political settlement, and the formation of a governing executive in which the DUP shared power with Sinn Féin.

Northern Ireland's peace process illustrates how co-operation need not require trust, but may struggle to proceed without it. We are right to warm to the word 'trust'. It denotes a condition that people desire in the same way that they desire love, and which is a fundamental element in any vision of a good society. It can readily be shown to improve social and political interactions, and to enhance the quality of everyday life. But it is not always good or necessary. Not only may trust be placed in error, as parents point out when they warn children not to take friendliness from strangers at face value, but it may also provide support for parties that do bad things. A Northern Irish pensioner may trust the paramilitaries who dominate her neighbourhood to deal with the young tearaways who are harassing her, but their trust-confirming response is likely to be brutal and their larger effect on society malign. To understand the value of trust, it is necessary to develop a sense of where it belongs and where it does not belong. We also need to develop an understanding of its relationship to other phenomena that are felt to be generally but not always desirable, such as cooperation and solidarity.

Theorists of trust have worked hard, particularly in the past twenty years or so, to develop precision in the understanding of trust and its relationship to allied phenomena. They have applied the theory of games to formal models and to experiments; they have conducted anthropological fieldwork; they have attempted to measure trust in public life; they have sought to clarify the concepts involved. The project is social science with an admixture of philosophy, applied mathematics, and influences from other fields, including biology.

The latter domain offers the opportunity to begin at the beginning. To address the fundamental problem of trust—how it can develop in a society of competitive individuals so richly endowed with the means to deceive or otherwise exploit each other—it is sensible to start with individuals and their interests, and to see where these may lead.

2

TRUST FROM THE BARREL OF A GUN

AN ANIMAL will often pursue its interests, as Richard Dawkins and John Krebs note in their text 'Animal Signals: Information or Manipulation?', by manipulating objects in its environment.[1] If the object is inanimate, the animal must apply physical force to make it move. If it is animate, the animal may be able to induce it to make the desired movements itself, by sending it signals. This form of manipulation is called communication.

A key feature of communication is what the biologist J. B. S. Haldane called 'the pronounced energetic efficiency of signalling: a small effort put into the signal typically elicits an energetically greater response.'[2] It takes a lot less energy to summon others than to drag them to where one wants them; it takes a lot less energy to send a request for another individual to go and find food than to go and find it oneself. Communication could be characterized, Dawkins and Krebs say, 'as a means by which one animal makes use of another animal's muscle power'. It can be said to occur 'when an animal, the actor, does something... to influence the sense organs of another animal, the reactor, so that the reactor's behaviour changes to the advantage of the actor'.[3]

Communication will tend to evolve because it serves the actor's interests: 'whether the reactor benefits or not is incidental'.[4] The first section of the text is headed 'The Cynical Gene', signalling that the text's intention is polemical. Dawkins and Krebs are posing a radical challenge to traditional views of animal behaviour, which had been inclined to assume that behaviour evolved for the good of species, rather than individuals, or genes. A traditional naturalist would tend to see communication between animals as a process harmonized by the greater good. It was not manipulation; it was an exchange of information that tended to be reliable, being in the mutual interests of the communicating parties. For Dawkins and Krebs, that was rose-tinted greater-goodism. Animals' signals are shots in the struggles between individuals. 'If information is shared at all it is likely to be false information.'[5]

This is the proper way to begin an analysis of interactions between agents. All that one can say at the outset is that individuals will tend to pursue their own interests. Cooperation cannot be assumed. Its emergence in a world of competing individuals is what needs to be explained.

Explanations are, however, rapidly forthcoming as this line of analysis is followed. Since communication takes at least two parties, one to send a signal and one to receive it, the interests of the receiver or reactor will be engaged. Reactors are likely to evolve a resistance to signals that harm their interests. That in turn may stimulate actors to evolve new behaviours that overcome the reactors' resistance; the reactors develop new countermeasures in turn, and an arms race is under way. This is likely to become a war about truth, in which the actor strives to assert and the reactor to test the reliability of the signals.

The most straightforward criterion for a signal's reliability is that it is hard to fake. Signals whose effect is to advertise an individual's strength or size should be of the kind, strong and sustained, that only a strong or large individual could produce. The more elaborate the advertisement, the more reliable an indication of quality it is likely to be. A male bower bird could surely not construct its display, a masterpiece of ordered extravagance, if it were in less than peak condition. Although receivers' reactions rarely drive signal costs to such extremes, by and large costly signals are more likely to be reliable than cheap ones. This is why the world of communication is so cacophonous.

While cost is the basic criterion of reliability, it is supplemented among animals with well-developed cognitive capacities by form. Animals capable of recognizing individuals and remembering their behaviour can assess signals in the light of the actors' previous actions. And some communication is reliable, as Krebs and Dawkins acknowledge in a second essay, because it is in the mutual interests of the senders and receivers. Mutual interest may relieve the parties of the obligation to produce and attend to costly signals; instead, they can engage in what Krebs and Dawkins call 'conspiratorial whispering', communication that can be quiet and concise because it is not under pressure to prove its reliability.[6] Human commerce offers an analogy in the contrast between the few quiet words needed to buy an item in a Western shop, with the price fixed and the customer's interests protected by layer upon layer of trading controls, and the florid protestations involved in haggling for goods in markets that lack a bureaucratic safety net.

The overall effect, as John Maynard Smith and David Harper point out in their book *Animal Signals*, is that most signals are reliable; their receivers would ignore them otherwise.[7] An analysis of behaviour that begins with an uncompromising and exclusive assumption of self-interest thus quite rapidly leads to an account of how self-interested behaviour produces general reliability in communication, and how mutual interests can create calm amid the general clamour.

It also testifies to the extent of the potential for honest communication by recognizing how this may arise between antagonists, even mortal ones. Often when a gazelle or antelope is being chased it will spring into the air, all four limbs straight. This action, known as 'stotting', looks dangerously whimsical. It slows the animal down and uses energy that might be vital to survive a long pursuit. But it makes sense as an example of the 'handicap principle' proposed by the biologist Amotz Zahavi.[8] By stotting the gazelle claims that it has resources to spare; that it can waste energy and still be likely to outrun any pursuers. The claim appears to be well founded, judging by field observations showing that the more gazelles stotted, the less likely they were to be killed by wild dogs. Stotting serves the interests of signaller and receiver: both avoid the costs of a long chase, and the signaller avoids the risk of death. That is transferred to the gazelles who are less capable of stotting.

Besides mocking the traditional assumption that animals act for the good of the species, this example illustrates that honesty does not require goodwill, and that acting in another's interest—in this case responding to the stotting signal by not pursuing the signaller—may fall a long way short of what is normally understood by cooperation. It is also an example of a game: a situation in which a player's optimal course of action depends on what other players do. Games are the basis of our understanding of how cooperation may arise, or not. Without understanding the evolution of cooperation it is not possible to understand the evolution of trust, even though the former can often be sustained without the latter.

At the hub of the question is the game known as the Prisoners' Dilemma. Its narrative cladding outlines a situation in which two prisoners accused of collaborating in a crime are held separately. Each is told that if he confesses but his partner remains silent, he will go free, while his partner will be given a lengthy sentence. Conversely, if he remains silent and his partner confesses, he will get the long sentence and his partner will be freed. If both confess, each will receive a sentence of medium length. If both remain silent, each will receive a short sentence.

Maintaining silence represents cooperation (with each other, not with their captors) while confession is what in game jargon is called defection (from a cooperative arrangement, whether explicit, implicit, or potential). Prisoners' Dilemmas are defined by an order of preferences in which defecting while the other player cooperates gives a better pay-off than mutual cooperation, which in turn gives a better pay-off than mutual defection, which is preferable to cooperating while the other player defects. Played once, as in the story of the captor's offer, a Prisoners' Dilemma will not result in cooperation if each simply reckons the possible outcomes. Each will calculate that he will do better to confess than to stay silent, so each will receive a longer sentence than if they had cooperated.

If the game is played over and over, however, players can make their choices in the light of each other's previous actions. In the early 1970s the political scientist Robert Axelrod staged a computer tournament in which different strategies competed in 'iterated' Prisoners' Dilemmas. The winning strategy was the simple tit-for-tat: cooperate in the first round, and thereafter do what the other player did in the previous round. One cooperative move thus prompts a cooperative response; any defection immediately triggers a defection in response. With the possibility of punishment, cooperation may be sustained indefinitely.

The sequences of code that process Prisoners' Dilemmas on computers are not conscious and hold no beliefs or expectations about the future actions of other agents. For humans, however, it is almost impossible not to form such expectations. Robert Axelrod saw the algebraic preferences of a Prisoners' Dilemma when he read the British sociologist Tony Ashworth's account of the 'live and let live' systems that evolved on many stretches of the Western Front during the First World War. Both analysts mark how these systems were based upon the developments of beliefs on each side of the lines about the intentions and likely actions of the soldiers on the other side. As the beliefs strengthened, they became trust.

The action in the western theatre of the First World War began as expected: a mobile conflict of a familiar kind, in which cavalry could play a meaningful role. Soon, however, a stalemate developed and each side dug in to face each other from trench lines; they remained entrenched for the rest of the war. In some sectors the men might be several hundred yards apart, but often they were close enough to shout at each other. Sometimes the enemy was only a grenade's throw away. The soldiers in the front lines were a lot closer to each other than they were to their own high commands, and in many cases this became true of their sympathies as well as of their physical distance. Martial ardour increased with distance from the front and reached its maximum virulence in the home country. 'There is only one temperature for women of the British race and that is white heat,' declared a 'Little Mother' in a letter that became a mass-selling tract. 'We women pass on the human ammunition of "only sons" to fill up the gaps.'[9]

With the forces immobilized, the high command hoped that its war aims could be furthered through attrition. They wanted their soldiers to wear the enemy down by inflicting casualties steadily. The men in the front line were therefore obliged to use their weapons or face punishment. They could exercise a degree of choice, however, over how they used their weapons. Some of them, especially members of crack regiments with a reputation for combativeness to uphold, pursued preferences identical to those of the high command. Their sectors of the line were infernal regions in which bands of raiders descended upon enemy lines at night with bombs and bayonets, and a man who raised his head even for a moment above the parapet would be killed as surely and instantly as a mouse that puts its head into a trap. In other sectors, however, soldiers could walk around in full view of their opponents, or even relax with a book in the grass behind the lines, without fear of attack. These troops were mutually implementing preferences completely at odds with those of their high commands.

In Axelrod's analysis, the soldiers who decided to 'live and let live' had the choice between whether to shoot to kill or to shoot so as to avoid doing harm. Shooting to kill equates to defection, shooting not to counts as cooperation. The preferences are in the Prisoners' Dilemma order: for each unit, defecting while the enemy unit cooperates is preferable to mutual cooperation, which is preferable to mutual defection, which is preferable to cooperating when the other side defects; but in repeated exchanges, both sides prefer mutual cooperation to alternating rounds of cooperation and defection. The latter preference was critical. Where units faced each other for extended periods, the cooperative possibilities of an iterated Prisoners' Dilemma became open to them.

Soldiers probably began to live and let live without really meaning to. If the units on each side ate their meals at the same time, each would appreciate the pause in hostilities, and might therefore be inclined to allow the enemy to eat his breakfast in peace. This might be no more than self-interest at first, but simply observing the scene on the other side of the line might plant the seeds of sympathy. Men saw their opposite numbers eating their meals or mending their defences, mirroring their own daily routines. They came to see the enemy soldiers as people with lives and a predicament similar to their own; people with whom they had much more in common than with their distant commanders. 'Tommy' and 'Jerry' became something like neighbours.

Once each side had begun to let the other live, some soldiers took initiatives to make explicit arrangements limiting aggression. The most famous instances were the now-legendary Christmas truces of 1914, but open fraternization of that kind was easily suppressed by senior commanders. The Christmas truces were the tip of an iceberg of truce arrangements that were mostly tacit and covert. They were sustained for the most part without explicit communication: by inactivity, and then, when pressure from the high command forced the troops to look busy, by ritualized fire. Shells would be fired at the same time and at the same target every day. On one section of the front soldiers understood that they were only at risk from enemy fire between two and four in the afternoon.[10] When senior officers' supervision or the distance between the lines inhibited verbal contact, the troops actually came to rely upon their enemy's fire for the information necessary to sustain truces. 'The constant flow of ritualised violence allowed each antagonist to know the other's current mood as well as to predict, with some confidence, his future behaviour,' Ashworth observes.[11]

For the system to work, it was necessary that each side believed in the other's ability to retaliate effectively if the rules of non-engagement were broken. One group of German snipers would fire at a derelict cottage so as to cut a circular hole in the wall: Axelrod suggests that such shows of marksmanship 'helped police the system by showing that restraint was not due to weakness.'[12]

There were times when one side's restraint did fail and provoked retaliation. To a reasonable approximation this was a tit-for-tat process. Battalion A would cooperate in a tacit truce with Battalion B until B defected, by performing unacceptably hostile acts, and would then itself defect by retaliating. Both the decision to retaliate and the extent of the reprisals could, however, depart widely from the simple rules of tit-for-tat. Retaliation was often a multiple of an eye for an eye. A British soldier noted that he and his comrades avoided using rifle grenades because the Germans fired three back for every one shot at them. But by the nature of the situation, in which shots and explosions were inevitable, the system depended on a degree of tolerance. If retaliation was automatic, truces would not survive: a shot would be answered with a shot, which would be answered in turn by a shot, and so on. Tolerance might be roughly algorithmic: a certain number of shells per day could be considered acceptable, for instance. It was also modulated by intuition and mood. Sometimes saying sorry might help. A British officer recalled how, after a salvo of shells from the German side, 'a brave German got on to his parapet and shouted out "We are very sorry about that; we hope no one was hurt. It is not our fault, it is that damned Prussian artillery." '[13]

This apology said even more about the system than about the salvo. The German expressed concern for the well being of the British troops, and was further concerned that the British would think that his group were responsible for what amounted to a breach of an ethical code. The game of non-aggression, having started from plain self-interest, had developed dimensions of sympathy and morality.

In this case the sympathies were evidently not universal. The German and his fellows were Saxons, who were antagonistic to Prussians and sympathetic to the 'Anglo-Saxons' of the British army. One departing Saxon unit urged the British to give their Prussian replacements hell. Ethnic sentiments could thus cut across political and military lines. So, perhaps, might sympathies based on class. This was the phase of industrial history when men in their masses developed an emphatic sense of being working class: the time by which, as the historian Eric Hobsbawm has observed, countless working men suddenly adopted a voluntary uniform of cloth cap and scarf, as if to assert the culture and interests they shared.[14]

German sympathies for British soldiers were also rooted in the extensive contact between the two nations before the war. Many Germans had worked in Britain, creating a legacy played on by the joke that when a British soldier called out 'Waiter!', fifty German voices answered 'Coming, sir!' Sometimes Germans would call out to ask about English football results. One ingenious group fired across a rifle grenade from which they had removed the explosive, replacing it with a message offering to deliver newspapers by the same means. Popular music could also bridge no-man's-land. Some machine-gunners would lighten the mood of the proceedings by firing bursts to the rhythms of popular songs. A German gunner's rendition of the line 'Steady, boys, here comes a copper' was completed by gunners on both sides, firing two rounds to add the final 'Good night!'

All these signals, from supposed ethnic affinities to mutual sporting interests, served to increase the sympathies of men on each side for each other. Happy Harry, one of a number of 'musical' German machine-gunners with recognizable styles, 'came to be quite liked, even as he was much admired' by his notional British targets.[15]

The more one feels one has in common with someone, the more confident one is likely to be about their behaviour. Indications of shared social and cultural experience would have added to the empirical confidence the soldiers developed by observing the ways in which their opposite numbers avoided doing them harm. And these perceptions were rooted in a fertile soil of habit. Tacit truces had arisen from everyday habits such as that of regular mealtimes. They were sustained by a kind of familiarity not unlike that which creates a sense of neighbourhood in a peaceful locality. One British soldier told a brigadier that there was an 'elderly gentleman' with a long beard who often showed himself above the German parapet. The brigadier demanded to know why the soldier had not shot the German. 'Why, Lor' bless you sir,' replied the soldier, ' 'e's never done me no harm.'[16] Familiarity permitted the application of a moral code based on reciprocity. The soldier recognized the German as an individual, part of his neighbourhood, and thus reduced the game to a two-hander, to be decided solely by how he and the elderly German gentleman treated each other.

Two soldiers did not make a truce, though. The signals had to be the right size. They had to be big enough to be noticed by the men on the other side of line, but not so big that they were noticed by the commanders behind the lines. A decision by an individual soldier to refrain from deliberate harm would not send a signal large enough to produce a response from the other side. Also, of course, it would be worse than useless if the decision was not upheld by the soldier's comrades. The bearded German would not have raised himself above the parapet if he had not been confident that all the British soldiers opposite would refrain from shooting at him. Truces were typically achieved between battalions, units large enough to keep hostilities suspended over a significant length of the front, and small enough to get away with it.

To a degree, the British army itself was responsible for the state of collective mind that induced battalions to decide for themselves whether to fight. It was 'intensely tribal', observes the historian Dan Todman, 'in the sense that it rapidly developed strong regimental loyalties among its men, but this also encouraged them not to look far beyond the boundaries of their own battalion'. Todman is offering an explanation for the lack of evidence that soldiers blamed their most senior commanders for their predicament. Generals like Douglas Haig, he suggests, were simply too remote from the front-line soldiers' experience to figure in their landscapes. Such indifference, arising from the narrow horizon of the battalion, may also help to explain the truces. A soldier with a mind to suspend hostilities was inclined neither to blame the generals nor to pursue the generals' aims at the expense of his own and his comrades'.[17]

It might have been possible to develop tacit truces on an impersonal, algorithmic basis, of a kind that could be accurately rather than approximately simulated on a computer. Neither words nor trust are needed for cooperation, Axelrod judges: reciprocity is enough. But the live-and-let-live systems of the Western Front were not merely brittle ceasefires. They were based not just upon a common interest in survival but also upon common cultural interests, familiarity, routine, an awareness of individuals, friendly gestures, and shared understanding about permissible behaviour—in sum, much the same set of factors as serve to create a sense of neighbourhood in peacetime. Live and let live was more than cooperation. It was infused with sympathy and it was illuminated by trust. Battalion A came to expect that the actions of Battalion B, facing them, would take A's interests into account. These expectations arose not just from the judgement that B's interests would be served by not provoking A into retaliation, but from an atmosphere of sympathy generated by a range of circumstances and confidence-building measures. The outcome was that A and B trusted each other not to attack in earnest unless forced to do so by their high commands.

Eventually, the British high command did find a way to enforce aggression and bring the tacit truces to an end. 'Live and let live' depended upon meaningful and consistent relationships, both across the lines and among the soldiers within each unit, who had to accept and sustain the informal truce rules. It could not survive the policy of trench raids instituted by the high command, involving up to two hundred men, and a complex division of military labour. These formations were too heterogeneous for any conspiratorial attempts to fake the aggression, and their results too damaging for the enemy to sustain without retaliation. Turning quiet sectors into zones of unpredictable menace, they disrupted the relationships that had developed in the front lines, destroying trust and creating distrust.

It must be admitted that for Axelrod this is not the most important consideration. He maintains that the 'foundation of cooperation is not really trust, but the durability of the relationship'.[18] For cooperation to thrive, he judges, the key requisites are that cooperative actions are reciprocated, and that the future casts a long shadow. By 'the shadow of the future' he means the parties' perceptions of the importance of future actions.[19] The more that A is concerned about B's next move, the more inclined A will be to cooperate in the current round of the game. Yet although trust may not be the foundation of cooperation or a key requisite for it, its salience in the truce system is remarkable.

Intuitively it is easy to imagine its subjective importance to the men involved: that for many of them, feelings of trust might have been at the heart of what it was like to be in a truce. It is also possible to discern a special role for trust in promoting the tolerance necessary to maintaining truces under disruptive conditions. Attacks that breached the understood terms of truces would inevitably occur. If the attacked side A felt that the attacking side B lacked commitment to cooperation, that B would be ready to pursue its interests at the expense of A's if the opportunity presented itself, A would be likely to perceive any such attack as a calculated breach of the agreement and retaliate in kind. But if the men on the receiving end had come to trust that the men on the other side would not attack them in earnest unless forced to do so, they would be more inclined to hold their return fire and to entertain positive interpretations of the event—to accept, for example, that it was perpetrated not by the other party to the agreement but by 'that damned Prussian artillery'.

Taking the long view of the evolution of cooperative behaviour, trust is a recent and inessential development. As long as B scratches A's back if A scratches B's, A and B can co-operate without trust, or the brains with which to trust, or without backs, for that matter. Trust arises relatively late in the evolutionary day, when animals become able to make predictions based on experience about the behaviour of others. It is not uniquely human. 'We have no trouble recognizing the difference between a trustful or distrustful dog,' observes the primatologist Frans de Waal, 'and we know how long it can take to turn the latter into the former.'[20]

Among humans, too, cooperation may be initiated and sustained without trust. But once trust becomes possible, it sustains interactions that would otherwise collapse, enhances the quality of cooperation, and threads the social fabric together. It is a prized sentiment whose absence is unthinkable in many contexts, and which is sought in contexts where reason might not find it to be strictly necessary. When our passions for a sentiment such as this run so high, our instincts are probably right. We value trust instinctively because it works for us, and has worked for our ancestors, in ways both familiar and beyond our grasp.

3

REASON TO BELIEVE

THE CONUNDRUM of language is that an animal clever enough to speak seems too clever to be believed. Many scientists believe that humans' unique intelligence may be Machiavellian in origin, that it ran away with itself under the selective pressure of life in groups of clever, calculating individuals.[1] And no evolutionary innovation has done more to increase the potential for deception than language. Words are cheap signals that slash the cost of manipulating others.

As John Maynard Smith and David Harper observe, the central problem for an evolutionist interested in animal signals is why they are reliable. Language poses this question in its most intriguing form. How could a signalling system that facilitates deception become established in an intelligent social species? As hominids began to speak, how did they find reason to believe each other?

In the dialectic sketched by Dawkins and Krebs, attempts to manipulate other animals using signals will meet with resistance on the part of those receiving the signals, which in turn will stimulate increased efforts by the senders to make their signals convincing. When sending, animals will be driven to invest in costlier signals, such as elaborate and physically demanding courtship displays; when receiving, they will be driven up the ramp of scepticism. Among intelligent social animals, living amidst intense signal traffic that must demand much of their attention, levels of scepticism should be particularly high. Yet the most intelligent and social animals of all have become defined by a communication system based on cheap signals.

It is possible to imagine that a system of words and grammar could be established without having to tackle a highly developed instinctive scepticism, if all that was said using this language could be verified in principle and with little difficulty in practice. Speech would be largely about the concrete and the immediate. Statements such as 'these berries are ripe' would soon be proved true or false. The difficulties arise, however, when we try to envisage the development of a language system used for statements about dispositions—'he likes you'—and other mental states, or abstractions, such as purity or luck. It is in speech of this kind, often unverifiable by its nature, that the qualities we feel to be most characteristically human inhere. Speech that speaks to our deeper feelings demands trust.

One account, offered by the anthropologist Chris Knight, proposes that ritual and language co-evolved. Rituals are costly in terms of adornment, activity, time, discomfort, and often pain. When individuals undergo rituals to join a group, they pay a price that has also been exacted from the other members. Having sent these costly signals, they are accepted as reliable, and so are the cheap signals of their speech. In the evolution of language, ritual guaranteed words as a bank guarantees banknotes.[2]

Cheap signals may also be backed by expensive ones at the point of delivery. Emotional displays tend to be costly and hard for most people, actors excepted, to fake. A shouting match is more costly in energy, and often in unintended consequences, than conspiratorial whispering. The shouts will probably be accompanied by other physiologically demanding responses, such as increased heart rate and reallocation of blood supplies within the body. Public speaking, which must be both audible and persuasive to a large and sometimes disparate group of listeners, likewise demands significant physical effort as well as investment in oratorical technique. The cost of words themselves, and their value, can be increased by crafting prose or verse so as to indicate the effort and resources invested in them.

Other mechanisms to promote reliability may be identified. The most basic, far from exclusive to humans, is the ability to recognize individuals and to remember their actions. Humans can extend this capacity by sharing information about individuals' behaviour to produce reputation. The information shared may not always be reliable, and reputations are not always deserved, but the value of being able to develop expectations on the basis of others' experience is not to be underestimated.

Humans also have the decisive ability to standardize behaviour by instituting norms and moral codes. Once rules have been established to govern conduct with respect to others, the range of actions that individuals are likely to encounter will be narrowed. Life will become more orderly and more predictable. In assessing B's trustworthiness, A can refer to a codified list of prescriptions that should, if B adheres to them, determine how B will act in respect of A's interests. This will also enhance the value to A of information about how B behaves with respect to others, since rules harmonize the treatment of persons. Compliance will require that, in respect of the rules, B treats A the same as B treats other persons—or other persons belonging to the same social category, at any rate. If A knows that B deals with such persons in a trustworthy manner, A can be confident that B is worthy of A's trust.

Moral codes are not grids imposed upon moral chaos from above. They take moral sentiments, the instinctive modes of moral thought and feeling that distinguish us from robots, and turn them into systems. The establishment of a moral code is not like razing a wilderness and building a city upon the cleared ground. Rather it is like the work of traditional farmers, who clear woodland but spare stands or rows of trees to define their new territories, and introduce plants that will flourish in the soil. There are already underlying patterns in the way that people feel about how they should treat each other. People seem to be attuned to fairness and to react strongly when they perceive unfairness, which they are often anxious to bring to the attention of others.

Some primatologists argue that the sense of fairness is far older than the human species, having evolved long enough ago to be observable in apes and monkeys. Sarah Brosnan, Hillary Schiff, and Frans de Waal conducted experiments with chimpanzees and capuchin monkeys in which the animals were paired and given either cucumber slices or grapes in exchange for tokens. Like children, chimps and capuchins prefer sweet food to cucumbers, and animals of both species would refuse cucumber rewards if they saw that their partner was rewarded with grapes. They acted as though they felt they had been treated unfairly.

Experiments on humans also suggest the existence of a deep-seated sense of fairness. In the Ultimatum Game, player A is given a sum of money and is invited to offer some of it to player B. If B refuses the offer, A must return the entire sum to the experimenters, and so neither player emerges from the game with any reward. Reason would dictate that A should offer a penny and that B should accept it, because even a penny is better than nothing. Respect outweighs reason, though. Players in the role of A tend to offer something close to an even split; and when they offer a meagre share they are often refused, even if the actual sum is not trivial. Thus donors usually act with a tendency towards fairness, encouraged by recipients' readiness to punish perceived unfairness.

The Trust Game also involves a donor A who is given a sum of money and invited to pass some of it on to a responder B, whereupon the transferred sum will be trebled. B then has the opportunity to give some of the money back to A. Taking a myopically rational view, B has no incentive to give anything back: realizing this, A should hold on to the entire sum. If they cooperate, however, both A and B can end up better off than they started. In practice, experimenters found that nearly all players in role A sent money to B, on average about half the sum they had been given. A third of players in role B returned more than they had been sent. Nearly all A's appeared to trust the B players to reciprocate. While the majority of A's were disappointed, the proportion of B's who cooperated was sizeable—and impressive by comparison with the rationally predicted figure of zero.[3]

The way that individuals respond to questions of fairness varies in different circumstances. Chimpanzees may be more inclined to tolerate inferior rewards if they have close relationships with their partners in the experiment.[4] Among humans, culture may affect expectations. While people generally offer close to half of their cash allocation in most places where the Ultimatum Game has been played, from Arizona to Indonesia, Machiguenga people in the Peruvian Amazon offered only a quarter, and accepted almost any low offer. The Machiguenga seem to feel no obligation to make the near-equal division that elsewhere would be considered a fair offer, nor any resentment at receiving offers that would be regarded in other cultures as unfair.[5] This does not imply that the Machiguenga lack a sense of fairness, or that such a sense is not a universal instinct, but it illustrates that different cultures may arrange the moral sentiment switches in different patterns.

Differences in the engagement of moral sentiments may vary in far more than quirks of detail. Some groups may apply moral considerations only among themselves, and consider all outsiders to be moral outlaws regardless of whether they have committed any offence. Within a society, values held by different groups may produce diametrically opposed sentiments of fairness. One of the clearest contemporary examples concerns homosexuality. Secular liberals are outraged by discrimination against homosexuals; conversely many people with traditional religious beliefs are outraged at the idea that homosexuality should not be discriminated against. Moral sentiments may be universal, but they are selectively directed. Their ability to harmonize behaviour and promote trust will depend upon the extent to which 'the benevolent affections', in the words of the nineteenth-century historian William Lecky, can expand the circle of their embrace from the family to 'first a class, then a nation, then a coalition of nations, then all humanity'.[6]

Lecky's geometry, placing the family at the centre of the circle, would be endorsed by evolutionary biologists a century later. They answered the question of why individuals should act in others' interests by reasoning that individuals share interests to the extent that they share genes; and added that reciprocity could sustain cooperation between unrelated individuals. In the scene that opened the first chapter above, shared genetic interests would form the foundation of co-operation, and hence trust, between mother and child. The transaction between the child and the shopkeeper would then be governed by the principle of reciprocity, requiring that the goods be paid for and the exchange conducted according to accepted standards of civility on both sides.

Considerations of genetic interest also help to explain why in that illustration it was more suitable to cast the parent as a mother rather than a father. The bonds between a child and its father can be as deep and strong as those between it and its mother. These days fathers are often to be found in kitchens too. But it is far easier for fathers to defect from parenting, or for defection to be imposed on them when relationships break down. Motherhood makes obligatory demands from the point of conception, at which fatherhood's required conditions are fulfilled. After she has carried the baby and given birth to it, the mother nurtures it to a degree of intimacy within which trust may be absolute. The father may well be sufficiently close, in his own way, to bring about a state of absolute trust; but on the other hand he may be absent from the scene altogether. The conditions for absolute trust are more often met in motherhood than in fatherhood.

Another asymmetry between parents that has far-reaching implications for trust is that of confidence in parenthood. A woman is almost always certain that she is the mother of her child; a man may need convincing that he is the father of a child for whom he is expected to provide. Mistrust in such cases is easily inflamed and hard to quell; the punishments it incites can be catastrophic and sometimes murderous. Men's explosive reactions to perceived sexual infidelity are unreasoning, but consistent with their reproductive interests. Their prevalence, even where social mores promote conciliation and deprecate male possessiveness, suggests a fundamental, instinctive concern that women should incorporate their partners' reproductive interests in their own. Women are also attuned to their reproductive interests, and may react passionately when they feel these to have been betrayed, but in most known societies they have not been in a position to insist that trust should be mutual.

People may also differ in the capacities that make trust possible. Some people tend to trust, others to mistrust. One might suppose that the trustful ones are gullible and the mistrustful ones are equipped with more sophisticated Machiavellian radar. Studies suggest, however, that the reverse is the case. The trustful ones are the better judges of character. The mistrustful regard others with suspicion because they have difficulty telling who can and cannot be trusted. Bad experiences discourage them from giving the benefit of their doubt, causing them to miss opportunities to interact with people or other agents they could safely trust.[7] Those better able to read the relevant signals can enjoy these opportunities; and they are generally trustful because they live among people who by and large are trustworthy. In some places, they might be as mistrustful as their less discerning fellows.

Trust thus develops, or fails to, in environments rooted in nature and extensively modified by human activity. To understand trust it is necessary to acknowledge the fundamentals of cooperation: the genetic interests shared by kin, and the power of reciprocity to sustain mutually beneficial interactions. It is reasonable to suppose that humans, like other animals, are guided in their actions by psychological dispositions or mechanisms—instincts, to use a traditional term—that have evolved in response to persistent features of their social environment, such as signals that must be assessed for reliability.

As the story proceeds from schematic self-interested agents, reacting like ions in a test-tube, to self-conscious individuals with the capacity to form expectations about each other, human artifice increasingly intervenes to complicate the picture. We now need to consider some of the many elaborate devices, such as contracts and codes of conduct, that humans have invented to promote trust or to substitute for it. But before doing so, we should turn back to a critical problem that has already appeared more than once without being properly examined, and acknowledge that some questions of trust demand a philosophical treatment.

In the Prisoners' Dilemma, rational analysis leads each prisoner to conclude that he should defect from the implicit alliance with the other prisoner. Prisoner A assumes that B will do what is best for B, works out that B would do best by defecting, calculates his own best move in that event, and concludes that he too must defect. In the Trust Game, similar reasoning would tell player A, who receives a sum of money and must decide whether to give some of it to B, that B would do best by keeping all the money B receives. If A were to assume that B would indeed keep the entire sum, and A was similarly determined to act according to self-interest, A would refuse to pass any money to B and so the game would never be played. This reasoning process, in which a player calculates what to do by working back from what would be best for the other player to do on the last move of the game, is called backward induction.

As the Prisoners' Dilemma shows, backward induction can present a formidable obstacle to cooperation. But the classic Prisoners' Dilemma seems designed to present human cooperation in the most pessimistic light (both in the logic of the game and its obscure moral topography, in which telling the truth appears to count as cheating). In practice, A players in the Trust Game defy the logic of backward induction, and are sometimes rewarded for it.

For the philosopher Martin Hollis, the problem is not whether people do in practice trust each other, but whether it is rational for them to do so.[8] In his book *Trust within Reason*, he sets out a whole series of backward-induction hurdles in a model he calls the Enlightenment Trail and which game theorists would call a Centipede. Two friends, Adam and Eve, set out on a walk along the Trail, which ascends through sunny uplands and is punctuated by six pubs. Adam and Eve agree that they will stop for a drink at one of these hostelries, and go no further. Neither cares for the first pub on the way, The Rational Choice. On a scale of 1 to 5, where 5 is best, Adam would give it 1 and Eve would give it 0. Nor are they keen on the next possible stop, The Social Contract, which Adam ranks 0 and Eve 2. Next is The Foole, 3 on Adam's ranking and 1 on Eve's; and after that The Sensible Knave, ranked 2 by Adam and 4 by Eve. The trail then goes on to The Extra Trick, which gets the maximum 5 from Adam and 4 from Eve, and finally reaches The Triumph of Reason, which Eve rates highest and Adam gives 4.

Regrettably, they never get to the end of the trail and The Triumph of Reason; not if they proceed by myopic self-interest, at any rate. If they combined their preferences by adding their scores together, they would get as far as The Extra Trick, bringing the walk to a convivial end at an inn they both like well. But if instead they compare each other's preferences, the slight difference between their scores for each pub will cause the entire exercise to unravel. Adam would go no further than The Extra Trick, because he prefers it to The Triumph of Reason; but they would not get that far, because Eve prefers The Sensible Knave; which they would not reach because Adam prefers The Foole; which Eve would avoid by opting for The Social Contract before it. As Adam likes The Social Contract even less than he likes The Rational Choice, he declines to go beyond the very first pub. Instead of an agreeable stroll through an uplifting landscape to The Triumph of Reason, which both of them like, they don't get beyond The Rational Choice, which appeals to neither of them.[9]

This sorry outcome emerges because Adam and Eve determine their preferences solely by how they rate each pub, as if walking the trail alone. They do not take their relationship with each other into account, as friends or companions normally do. The more Adam likes Eve, the more he will factor her preferences into his, especially if he wants her to like him more than she already does. Each may also feel the need to consider the other's preferences because their own preferences include acting towards others according to norms of fairness and sympathy. Adam will want to be kind to her because it is important to him. They might get further along the trail if their preferences were influenced by such considerations. But that would not be the same as adding their preferences together, to consider themselves as a pair or a unit. They would still be acting according to their individual interests. By excluding factors like these that make it look otherwise, thereby restricting preferences to those which would apply if each were walking alone, Hollis construes Adam and Eve as the kind of atomic individuals imagined by economists and biologists.

Individuals like these will always be looking out for a bargain, and they will always take it if they get the chance. This precludes the common sense solution that Adam and Eve should discuss their preferences rationally and agree on a choice that will be suboptimal for at least one of them. Bound to the wheel of narrow self-interest, they cannot keep a promise if a better opportunity comes into view. The personal best is the enemy of the common good.

Under these conditions social life seems impossible, for it 'depends on trust, especially on trust that promises be kept'.[10] Hollis quotes Friedrich Nietzsche's reflection upon the idea that an animal clever enough to speak is too clever to be believed: 'To breed an animal capable of promising—isn't that just the paradoxical task which nature has set herself with mankind, the peculiar problem of mankind?'[11]

Would Adam and Eve get any further with other thinkers as their guides? Hollis first considers Thomas Hobbes, and the power of fear. In the aftermath of the English Civil War, Hobbes argued that agreements between individuals must be enforced by the fear of punishment, threatened by the 'Leviathan' state that must be raised up above the people to impose peace and order upon them. To escape the state of nature, 'a war of everyman against everyman' in which life is 'solitary, poore, nasty, brutish and short', people must band together and erect a higher power that will use whatever force is necessary to suppress the inherent conflicts between them. For example, 'the early inhabitants of the American

Wild West, finding their lives nasty, brutish and short, might agree to appoint a sheriff for their mutual benefit.'[12] Leviathan engenders trust through fear. That gets Adam and Eve beyond The Rational Choice and on to The Social Contract.

A contract based on fear is a shaky one, though, depending as it does on the efficiency of the coercive power. If contracting parties think they can defect and get away with it, they often will. In *Leviathan* Hobbes introduces a character to point this out, but dismisses him as a 'Foole' for not grasping that justice and other moral virtues are divinely ordained laws of nature. Even in societies that retain a dominant idea of divine law, however, 'fooles' seem to abound. Nor do open modern societies countenance the kind of sovereign, handed absolute power by the people as though he could be absolutely trusted with it, in whom Hobbes invests his hopes for order.

Hollis allows that reputation may count for something, but it will not guarantee trustworthiness, because sometimes being honest will pay less than being wrongly thought honest. 'This is a serious matter in a shifting modern society, where people are opaque to one another,' he warns.[13] Indeed—but its serious consequences are incurred by the dishonest as well as the honest, which is an incentive to honourable behaviour. The economist Robert H. Frank suggests that this is a good, self-interested reason to leave appropriate tips in restaurants that one will never visit again. Tipping in restaurants to which one may return is a kind of investment, which should provide a return in the form of good service on subsequent visits. Tipping in restaurants when there is no risk of punishment, in the form of subsequent poor service, is a more profound kind of personal investment. By resisting the temptation to cheat, one maintains the consistency of one's reciprocal responses. Reciprocity in one-off transactions promotes reciprocity, and therefore trustworthiness, in relationships that are longer lasting and more meaningful. It keeps the capacity for commitment topped up. Cheat in distant towns, and it becomes harder to sustain the emotions that motivate honest behaviour back home. Those around may notice the change in disposition, and raise their guards accordingly.[14]

Things are simpler on the Enlightenment Trail, however, and although a social contract may take Adam and Eve on to The Foole, they will get no further because there is no power above them to stop either defecting from their agreement. Could sympathy succeed where fear falls short? Hollis turns to David Hume, whose 'eighteenth-century view of our motivating moral sentiments yields a wider and more genial list of passions' than Hobbes's, 'notably a natural sympathy for our fellows'.[15] Natural sympathy goes only so far, though. It is naturally partial, its influence strongest within the family and diminishing thereafter with distance from the centre of the circle of benevolent affections. Hume's individuals may be more rounded than Hobbes's, but they still need confidence that others will not take advantage of the better side of their nature. Rules of property and justice provide 'an artificial remedy for a natural partiality'.[16]

On the Enlightenment Trail, these might be sufficient to take Adam and Eve, their mutual sympathies now modulating their preferences, past The Foole. Having come that far, though, they would face the challenge of The Sensible Knave.

This character, according to Hume, may now and again be tempted to think that 'an act of iniquity or infidelity will make a considerable addition to his fortune, without causing any considerable breach in the social union and confederacy.'[17] Hume admits that the Sensible Knave has a point. And the Knave identifies a general problem in human relations. The most cooperative society will still be vulnerable to a degree of cheating. Indeed, a society's basic stability may encourage a minority current of cheating, since the consequences of occasional defection should not be catastrophic.

Adam will be keen to pass by The Sensible Knave, since the next inn on the Trail is his favourite, The Extra Trick. The difficult part is to induce him to carry on past to The Triumph of Reason. Immanuel Kant would insist that promises must be kept regardless of pay-offs. Kantians stick unswervingly to their duty, making them trustworthy. If Adam is a true Kantian, Eve can be sure he will keep his promise, and together they will march to the end of the Trail. The problem is solved.

It is not solved to the satisfaction of any real human being, though. Who wants to have a drink with somebody who has only gone to the pub out of a sense of duty? If trust in Eve's case is her expectation that Adam's actions will incorporate her interests into his own, she will be left feeling that Adam will uphold the letter of her preferences but not the spirit of her interests. Adam will do the right thing because it is right, not because he empathizes with her and wants her to be happy. Kantians can be relied upon, but their unconditional devotion to duty differs in quality from trust, which is conditional outside the special bond between a parent and a young child.

'Their morality has to be too high-minded to serve as the bond of society,' decides Hollis, who has noted the philosopher John Locke's observation that the bond of society is trust.[18]

To develop this bond, the members of a society must come to think of themselves as a collectivity. This must be more than the sum of individual preferences. If Adam and Eve add their preference scores together, they will hike past the pubs that they like little or not at all, but they will still have to find a way to decide between the last two pubs, for which their individual preferences differ although their aggregate scores are equal.

Instead, interacting parties must see themselves as a network of exchange. The basis of exchange is reciprocity; the existence of a network allows reciprocity to become fluid. A can scratch B's back without trusting B to scratch A's, if A is confident that a C, D, or other initial will come along to do the scratching. Direct reciprocity is replaced by indirect reciprocity, which has the potential to stimulate and sustain cooperation across networks of strangers.

Perhaps the most celebrated example is the donation of blood within Britain's National Health Service. The blood is given without payment and received without direct payment, health services being funded primarily by taxes. Donors must find the time, steel themselves against the insertion of a spiked tube into one of their veins, and yield a pint of their blood. 'A pint? That's very nearly an armful!' protested the comic actor Tony Hancock, in a fondly-remembered sketch. He may have been medically inaccurate, but his outburst affirmed the non-trivial subjective cost of the experience.

Asked why they should spare their time and a pint of their blood, donors typically replied that they might need a blood transfusion themselves one day. Their responses construed the system as an indirectly reciprocal one, in which reciprocation would be provided at some uncertain future stage by unknown and unidentifiable strangers. The way they put it, they gave in case they later needed to receive. But they would not be getting their own blood back, nor would they be denied a transfusion if they had omitted to donate in the past. The potential return could be life-saving, though the only certain and immediate pay-off was the traditional biscuit and cup of tea provided after the procedure—a token that, in a small but not entirely insignificant way, symbolically affirms the importance of reciprocity.

Both the donors' explanation and the token refreshment can be taken as implicit recognition that a sense of reciprocity is psychologically necessary even in systems of exchange that cannot balance what is given and what is received. The principle of indirect reciprocity seeds a system that allows people to behave altruistically, however they choose to rationalize it. And they trust the National Blood Service to provide them with blood should they need it, because they trust unknown others to continue to give blood for unknown others' benefit. Adam and Eve can drink to that in The Triumph of Reason.

4

IN GOD WE TRUST

THERE IS one glaring exception to the rule that absolute trust does not normally flourish beyond a child's early years and its relationship with its parents. What ceases to be possible with other people remains possible with a deity. The most successful and world-changing innovation in the history of trust must surely be the idea of an omnipresent, omnipotent god. Reason may struggle to lodge trust within it. Faith has no such problem.

Finite deities, such as those identified with mountains, sun or sky, tend to be unpredictable. Indeed, the creation of gods by ascribing conscious agency to inanimate objects may in large part have been an attempt to explain natural events that at the time were beyond any other explanation. In their classical form, deities were not just unpredictable but capricious and mischievous. The ancient Greeks would not have trusted their gods any further than they could have thrown them. With the advent of a single universal God, however, the deity's movements may remain mysterious, but they are absolutely trustworthy. Good and right are constituted in God, whose every action or inaction must be perfectly right. Believers should enjoy a profound trust in such a god—although that may be a secondary consideration in faiths whose central demand, as the derivation of the word Islam affirms, is submission and acceptance.

Trust demands two basic conditions, that the party being trusted will incorporate the trusting party's interests into its own, and that it is capable of the actions required. An all-powerful god fulfils the second condition absolutely. Its relationship with the first is more complicated, though. A morally perfect being cannot be expected to favour one individual's interests above another's without moral reason to do so—not that this stops footballers from crossing themselves as they run onto the pitch. Believers can trust God to do what is right, but not to do what they want.

A believer's interests can be squared with the deity's in various ways, with different interpretations and emphases emerging in different variants of religious tradition. One solution is to place spiritual interests above material interests. The condition of a person's soul is more important than the enjoyment of earthly pleasures, which may indeed be spiritually toxic: better then to act according to God's law than according to worldly desires. Following on from this is the principle that dutiful believers will get their reward in heaven. Like parents who act to protect their children's interests by denying them the sugary food they crave to excess, God knows his people's interests better than they do themselves. By living as God has instructed, denying themselves in the here and now, they stand to enjoy eternal bliss.

The alternative, as strenuously advertised by the activists of certain religious currents, is eternal torment. This necessarily renders trust in God conditional. The faithful may trust absolutely in God's love and mercy, but accept that they must meet certain standards of thought and action if they are to gain the Kingdom of Heaven. Their relationship with their god, the foundation of their lives, is contractual. Believers are given a body of texts instructing them as to how they are to do God's will, promising rewards for effort and indicating penalties for non-compliance. (Like secular contracts, such documents require interpretation by experts, raising secondary issues of trust.) In Christianity the principle of reciprocity animates the relationship between the believer and the deity. 'God so loved the world,' says the Gospel according to John, 'that he gave his only begotten Son, that whosoever believeth in him should not perish, but have everlasting life.'[1] Jesus died for your sins: in return you should believe in him and strive not to sin.

Believers may also trust that the Lord will provide for them in this world. In the early 1970s John Maynard Smith collaborated with an independent scholar called George Price in work that pioneered the application of game theory to the study of evolution. During this period Price underwent a deepening spiritual crisis, in which he felt compelled to adhere strictly to the injunctions that close the sixth chapter of the Gospel according to Matthew. This is the passage that urges its audience to devote itself to its heavenly Father, who will take care of its material needs: 'Therefore take no thought, saying, What shall we eat? or, What shall we drink? or, Wherewithal shall we be clothed?' In exchange for devotion, the Father's support is promised: 'But seek ye first the kingdom of God, and his righteousness; and all these things shall be added unto you.'[2] Having yet to receive a divinely guided cheque, and down to his last 15p, Price admitted to Maynard Smith that 'God's standards of what constitutes "disaster" are on a different scale from mine.'[3]

He told another collaborator, the evolutionary biologist Bill Hamilton, that he twice stopped taking thyroxine, a medication he needed, as a test of God's will. On one occasion the illness this induced led to a fall and a fracture; at the hospital he was given thyroxine without asking for it, and took this as a sign that God wanted him to continue his efforts to work for the poor while sharing their condition. Eventually he took his life, leaving a brief note explaining that because of the difficulties he had encountered in his charitable efforts and his worries about becoming a burden on his friends, he had decided the time had come to go to his Maker.[4]

It is impossible to be sure how he understood his relationship with his god by the time he made this final decision, but his earlier hazards with providence show how resistant to disappointment trust in a deity can be. At first he expected deliverance when almost penniless; later, no financial relief having materialized, he pointed out to Maynard Smith that he still had his Barclaycard. He did not appear to question the principle that if he did God's will, the Lord would provide. Instead, he seemed to interpret the non-arrival of provision as a sign that his chosen course might not in fact be what God wanted from him. To test this, he distorted the spirit of Matthew's Gospel by failing to take his medicine, one of the few necessities of his life whose provision, thanks to the

National Health Service, was assured. Matthew's reproaches to 'ye of little faith' were not addressed to him.

Price was an exceptional individual, but there was nothing exceptional in the manner of his trust in his Lord. When providence fails to meet expectations, it is normal for the faithful to identify an earthly cause. Something similar may be seen in the trust subjects or followers place in their temporal lords. They do not blame the King, the Leader, or Comrade Stalin for the injustices visited on them in his name. There are any number of subordinates in between who must be responsible: if only the victims could make contact with their lord, he would surely see that justice was done. The sovereign or leader is powerful, but not all-powerful and all-knowing; so the criterion of capability can be applied to explain the malfunction in the relationship.

Temporal powers have traditionally claimed divine sanction and demanded popular allegiance. All three parties are united in the oath, a simple device for promoting confidence and reliability; not necessarily the same as trust and trustworthiness, according to some definitions, but close kin. An oath is a conditional self-curse: the person swearing it pronounces a curse upon himself or herself, to be implemented should he or she act in certain ways. The structure may be clear and explicit, with punishment, condition, and punitive agent all specified—'May God strike me dead if I tell a lie!' Often the threat of punishment is left implicit in the reference to a higher power, which would be expected to impose penalties for transgression: 'I swear by Almighty God that the evidence I shall give will be the truth, the whole truth and nothing but the truth.' In other cases the promise may be general in nature. When citizens swear oaths of allegiance to their country, they pledge to incorporate their country's interests into their own.

The supernatural power of the oath has sustained it in the face of widespread deconsecration. Affirming that one will tell the truth is just not the same. It is something of an anachronism, however, in a world that has reduced its exposure to traditional authority. In making itself modern, the world has moved away from its reliance upon gods and customs as means of regulating behaviour. The reciprocal relationships between gods and persons are supplanted by reciprocal relationships between persons or organizations, agreed as contracts. These proliferate in societies where relationships are no longer considered sacred, as in holy matrimony, or divinely ordained, as in serfdom.

Contracts are substitutes for trust, and it is in their nature to elaborate themselves, specifying in increasingly baroque detail how relations between the parties will obtain in all manner of circumstances. A contract comes into being when an agreement regarded as binding in law or by custom is reached. The farmer and the journeyman agree on the payment for a day's labour and they shake on the deal. This is the contract in its plain, honest, and primitive form. Once it is written down, its innocence is lost and it brings modern relations into being. Primitive trading contracts still exist, but they are being replaced by numbered clauses as inexorably as the tractor replaced the plough-horse. Editors may still commission journalists to write articles by addressing them in much the same way as the farmer hailing the journeyman, but it is now normal for such verbal contracts to be supplemented by written terms. In higher education, a lecture may be commissioned with a contract that would take as long to read out as the lecture itself.

As traditional authority wears off, choices multiply. Halfway through the past century, deference to traditional authority still imposed a comprehensive standardization upon Western society, in which couples expected to marry for life and typically fulfilled their expectations even if it caused them misery, men of even the most advanced artistic leanings kept their hair above their collars, and shops were shut on Sundays. Then the West turned into a rock'n'roll culture, glorifying the fantasies and desires that had hitherto been repressed, in the name of self-fulfilment. Although professions of belief continued as loudly as ever in America, their connection with norms of behaviour were weakened. It became routine, if not de rigueur, for rappers to pose with guns, disrespect women, and thank God in their album sleeve notes.

With self-development at the heart of the personal project, individuals are encouraged to elaborate on their inner worlds and to draw eclectically from the outside world. Although in practice people tend to converge upon a relatively limited number of poles of attraction, sustaining a mass culture based on bestselling music, images, and styles of adornment, self-development leads to extensive social uncertainty. People settle on different combinations of norms, making it more difficult to form expectations about how they will behave without getting to know them, and thus making it more difficult to trust strangers or acquaintances.

In some respects it seems as though the attenuation of traditional authority has created a vacuum in social relations that is being filled by bureaucratic regulation. While codes of conduct proscribing sexual harassment or racist behaviour are primarily the result of political struggles to assert women's rights and racial equality, they can also be seen as mechanisms that have arisen to regulate social relations in the absence of generally accepted norms. Declarations of principle are now posted in public spaces: 'We work hard to provide you with an excellent service, and deserve to be treated with respect. Neither we nor our customers are here to be verbally or physically abused.'[5] These statements are part of the forward march of explicit contracts, colonizing the space of possible contingencies clause by clause. A 'Mutual Respect Statement' issued by the Women's and Children's Health Service of Western Australia rather elegantly asserts the right to 'defuse, deflect and respond appropriately to aggressive behaviour'.[6] With each such proclamation, from British post offices to an Australian children's hospital, the public gets the message that it can no longer be trusted. Those who don't need to be bound by contract to behave properly have their attention drawn to the threat from those who do, nudging their expectations of others in the direction of distrust.

As well as making it more difficult to anticipate the behaviour of strangers or acquaintances, the imperative to develop the self creates uncertainties about the behaviour of individuals over time. If people are constantly seeking new avenues of self-development, which may include dalliances with a range of religions or other belief systems, they may be constantly revising their behavioural norms. Or they may carry on behaving the same way, but generate uncertainty by maintaining a rapid turnover among the ideas they use to justify their actions. They are certainly likely to alter their understandings of their interests, which may in turn affect their readiness to incorporate the interests of others into their own. The will to self-development builds a tension into close relationships, which require work if they are to be more than a temporary coincidence of interests.

According to the British sociologist Anthony Giddens, the work required is mutual self-disclosure. Giddens sees the modern condition as one in which traditional authority no longer works, obliging people to make their own arrangements. That was the position the soldiers on the Western Front found themselves in, when they came to the conclusion that the traditional beliefs that had brought them there did not adequately inform their situation. They covertly defied their staff officers, who embodied traditional authority, and made their own arrangements with the soldiers across no-man's-land. Similarly, individuals living outside traditional authority today no longer move dutifully into allotted roles, but take their peers as they find them. As two modern individuals become intimate, they enter into what Giddens calls a 'pure relationship', one which exists only for what it brings to each partner.[7]

In the absence of external supports, the pure relationship has to bind itself together with trust, which 'can be mobilised only by a process of mutual disclosure'.[8] It is not enough to get to know each other's personalities by observation;

each must actively reveal their secrets. There is a premium on exclusivity: 'Intimacy means the disclosure of emotions and actions which the individual is unlikely to hold up to a wider public gaze. Indeed, the disclosure of what is kept from other people is one of the main psychological markers likely to call forth trust from the other and to be sought after in return.'[9]

When relationships are inherently fragile, unsupported externally, and vulnerable to the demands of the self, this looks rather like putting all one's eggs in one basket. Partners are induced to shift the distribution of trust towards each other and away from a wider circle of intimates, confidants, and friends, who might be able to offer a more reliable and productive range of responses to whatever is disclosed. On the other hand, we can learn in two minutes' leafing through a gossip magazine in a supermarket queue intimate details about celebrities, freely volunteered in interviews, that we might not learn about our closest friends in twenty years. From the perspective of the pure relationship, this amounts to a tacit acknowledgement on the celebrities' part that their relationships are transient; as does their replacement of trust by pre-nuptial agreements.

If the pure relationship, in the sense of relationships within or resembling marriages, sits a little uneasily with friendship, it is harder still to integrate with the needs of any children resulting from such relationships. Nor does it convey a sense of the very practical ways in which couples make relationships work effectively and fairly.[10] All the same, it seems to express something distinctive and important about the spirit in which people approach their closest relationships today. Crucially, it identifies tradition's replacement by trust as the binding of relationships; although it fails to recognize that trust develops from shared activities and responsibilities more than from disclosed secrets.

Giddens also identifies a distinctively modern role for trust in relation to what he calls 'disembedding mechanisms', which take social relations out of their local context and restructure them across indefinite spans of time and space.[11] These come in two forms, symbolic tokens and expert systems.

The prime and overwhelmingly dominant example of the symbolic token is money, which gives value indefinite range. It is, of course, the root of globalization. Buying a manufactured item such as an electrical gadget or a toy entails connecting two complex flows of value. One spans the retailer, the importer, the company that brands the product, and the firm, usually in the Far East, that actually makes it. The other connects the retailer, often via the internet, to the flows of revenue, capital accumulation, and investment from which the customer's wages are paid.

Although the nature of money has not changed with the advent of modernity, new services based on new technologies have multiplied the ways in which value can be transmitted. Credit cards extend the span of time between purchase and payment, a disembedding mechanism for which the customer may eventually pay a substantial charge. Public transport tickets are being replaced by smart cards; airlines and supermarkets have established their own private currencies with their customer 'reward' schemes. Internet payment systems such as

PayPal provide the means and the confidence for strangers to trade with each other in online marketplaces.

All these are implemented through systems built by experts and trusted by the general public. Few people understand the systems they rely on to any great depth, as the child whose trip to the shop opened this discussion would discover if he started asking adults questions about how traffic lights, banks, or even bakeries work. Nevertheless they drive, they get on planes, they pay money into banks, and they take what goes on under the bonnet or in the black box for granted. Their trust in expert systems is partly empirical: experience tells them that traffic lights change in the proper order, that airliners land safely at their destinations except in very rare cases, and that computers don't get sums wrong. It is also rooted in respect for authority. Trust in expert systems is trust in the experts who build the systems, underpinned by the belief that the knowledge they employ is objectively true.

If people's expectations about the performance of expert systems related solely to the electronics and the machinery, they might be better described as confidence. There is no question of believing that a computer can incorporate one's interests into its own, since computers do not have interests of their own. But the performance of the machines influences people's trust in the experts who design them, partly by demonstrating their competence and partly by offering empirical evidence for the objective truth of their knowledge.

Trust in experts and their systems thus occupies a position somewhere between trust in deities and trust in ordinary people. The well-worn comparison between scientists and priests applies in this context: both groups are respected because they are believed to have exclusive access to higher knowledge. Respect for scientific or technical authority disposes people to accept expert decisions, but they also need to see that those decisions work in practice before fully accepting them. With new technologies this may take time—half a century, in the case of aviation. When people joked that 'if God had meant us to fly, he would have given us wings,' they referred back to traditional authority because they had yet to be persuaded that aviation systems had become expert enough to trust.

More recently, people's suspicions about expert systems have centred on information and communication technologies. Their fears are not about dangers to life and limb, but to social relations and morality. The stereotypes that rose around the early Internet—before it stopped being an entity with a capital 'I' and settled down into the lower-case internet, an infrastructure system like the electricity grid—caricatured its developers as geeks, men who related to computers instead of people. By implication, a network created by men like these would not support healthy social relations. Their interests did not incorporate the interests of society as a whole.

Such suspicions were reinforced by the early adoption of the internet for sexual exploitation and abuse. People learned that the anonymity of online interactions could be exploited by individuals pretending to be something they were not. Virus authors exposed them to regular lessons about gratuitous malice and the hazards of assuming that strangers are trustworthy. As access became more widespread throughout workplaces and homes, however, people found that the internet could bring them closer to those they already knew and trusted; their relatives, friends, and intimates. Over time, most people who used the internet came to feel that its usefulness to them far outweighed its dangers.

Another factor that weighed heavily in their reckoning was their embrace of online shopping, with its cut prices and convenience. A major investment in expert security systems, protecting customers from thieves and fraudsters, was needed to give them the necessary confidence. They are now encouraged to help promote online trust themselves. Marketplace sites such as eBay and Amazon (which puts customers in contact with affiliated dealers as well as itself) ask users to rate their transactions. The ratings serve to promote confidence both in particular traders and in the ability of the system to identify unsatisfactory ones. A good rating becomes an essential business asset, providing a keen incentive for all traders to provide the best service they can. Ratings bring an element of community activity to the marketplace, encouraging individuals to give information about private transactions for public benefit, at a small cost to themselves in time and typing. Their credibility does, of course, require trust in the honesty of the people behind the usernames.

As the public began to explore the internet, many complained that the systems were not nearly user-friendly enough. Computers required too much technical knowledge. They were designed by experts for experts, who were unwilling to incorporate ordinary people's interests into their designs, possibly because their capacity to relate to ordinary people was limited. The complaints did not call the knowledge itself into question, though. As technical knowledge it is seen as objective and certain. Scientific knowledge is in a more ambiguous position. Technicians and technologists are assumed to agree with each other; scientists are frequently seen to differ. This is a constant source of disappointment to the public and the media.

The comparison between priests and scientists asserts that science has appropriated some of the authority traditionally monopolized by religion. Science has certainly replaced religion as the means for people who do not believe in the supernatural to understand the universe and their place in it. Among believers it has also assumed a higher explanatory profile, acquiring authority as the means to understand how divine creation unfolds according to the laws of nature. The difference between the two is that change is science's *raison d'etre*, whereas religions pride themselves that their truths stay the same. When people remark that 'the scientists don't agree among themselves', perhaps the note of complaint expresses disappointment that scientists are not more like priests.

It may also be a complaint about performance: that the experts should sort themselves out and decide what they think about epidemics, climate change, or the safety of medicines. The implication is that the scientists are untrustworthy, as they have failed to incorporate the public interest sufficiently into their actions. The assumption is that scientists can do the impossible. In one British opinion poll, 61 per cent of respondents looked to scientists to give 100 per cent guarantees about the safety of medicines. They demanded absolute trustworthiness and zero risk.[12]

Scientists may also be mistrusted because they are suspected of serving vested interests: companies, governments, 'Big Pharma' (the pharmaceutical industry, comprising a small number of very large players), or the medical establishment. When doubts about the safety of the measles-mumps-rubella (MMR) vaccine verged on panic in Britain in the early 2000s, distrust reached statistically measurable proportions. Vaccination rates dropped below the levels considered necessary to provide 'herd immunity' in certain areas, including London, and have not fully recovered since.[13]

The suggestion that the vaccine might induce autism was made in a single paper and was associated thereafter with a single researcher. It was taken up by parents who believed that vaccines had done terrible harm to their children, and by the media, especially those newspapers that have devoted themselves to the art of making the middle classes afraid. Ranged against them were the government, its scientists, the medical establishment, and a succession of studies that found no support for the claim.

An opinion survey taken at the height of the controversy indicated how the public ranked the various sources of information competing for their trust. Respondents were most likely to trust their doctor's advice. That is not surprising, since people's trust in their doctor is based on personal experience in the consulting room as well as the certificates on the wall—and if they mistrust their doctor, they will try to find themselves a different one. Nearly two-thirds said they would trust their doctor's advice a lot. Less than half said the same about scientists, and only a fifth expressed a great deal of trust in the government ministers, the Department of Health, and their elected representatives in Parliament. A mere 6 per cent said they would trust journalists a lot; though sales figures of about one daily newspaper for every four adults might suggest a more influential, if diffuse, role for the papers in unsettling the public mood.[14]

A failure to follow scientific reasoning may underlie some of the suspicions about the MMR vaccine, and in some cases the scare may have resonated with an undertone of hostility towards a medico-scientific system felt to be insufficiently responsive to human needs. As the philosopher Onora O'Neill observes, these days people place their trust in doctors and scientists 'erratically and with reservations'.[15] But the controversy was also about policy. The government was thought to be discounting the possible risks of the vaccine because of the importance it had placed on attaining targets, including vaccination levels, in its public health strategy. Its credibility was cast in a poor light by a history of controversial responses to public health crises, particularly the foot-and-mouth epidemic of 2001 and the emergence of bovine spongiform encephalitis (BSE) in beef cattle in the 1990s.

Democratic governments these days enjoy precious little of the deference, impressed on people by the weight of traditional authority, that curbed interference with their activities in earlier times. Politics, objective truth, and the public interest are only tenuously connected in the public's view of the scheme of things. During public health alarms it is perhaps not surprising that many people choose to believe the side the government is not on.

5

THROUGH THICK AND THIN

THE CHILD thanks the shopkeeper and walks out of the door, his loaf under his arm. As he leaves, a man steps across the threshold. He ignores the stock and does not make his business immediately clear, but the shopkeeper has a good idea what it is. The man has a proposition to make. There are threatening characters in the area, he says, who are liable to rob or wreck business premises such as the shop in which he is standing. For a regular payment, in cash, the organization which the man represents will protect the shop against such assaults.

To the general public, wise to the ways of organized crime through its viewing of television and movies, that kind of protection is a racket. The threatening characters are the man and his associates, who are demanding money with menaces. They are simply parasites, doing nothing but harm to those whose money they take.

This may be the greater part of the truth, but it leaves the question of why the idea of the protection racket has become so established and familiar. Why have criminal groups apparently been so concerned to suggest that they are offering a service? From the account given by the sociologist Diego

Gambetta, the answer appears to have roots in the policies of the Spanish Habsburg and Bourbon rulers who controlled southern Italy from the eighteenth century until the country's unification in 1861.[1] The watchword of each dynasty was divide and rule, the Bourbons taking particular pains to promote hatred between Neapolitans and Sicilians.

According to contemporary critics, the Spaniards destabilized commerce through capricious or corrupt financial measures and undermined the rule of law. Money was frequently counterfeit and bonds were false, discouraging trade and driving the cost of loans to exorbitant rates. The eighteenth-century political economist Paolo Mattia Doria even charged the Spanish rulers with destroying the relatively free and equal relations that had flourished between the sexes. After they had imposed their masculine code of honour, 'a conversation with a woman seems more like worship at a shrine than a discussion.'[2] Much of what they did, summarizes Gambetta, 'can be seen as the promotion and selective exploitation of distrust'.[3]

The result, he suggests, is a society which finds it difficult to trust itself, and is correspondingly unable to achieve the kind of cooperation that has made the northern part of Italy economically dynamic. In Palermo, the capital of Sicily, it could not even manage to establish a radio-taxi service like those which operated in all the other major cities of Italy. The normal modus operandi in such services is that when a customer requests a taxi, a dispatcher announces the customer's location, and drivers report how near they are; the nearest driver takes the job. This system requires a degree of trust between the drivers, since it is possible for them to lie about their locations. A service was launched in Palermo in 1981, but it foundered amid accusations of cheating. If any driver had a good day, his fellows would be sure that it was for a bad reason. 'In Sicily there is nothing as suspicious as luck,' Gambetta observes darkly.[4]

However profound the distrust it does not quite explain the failure of the system. In Naples and Milan, drivers could police the system by driving to the address claimed by another driver, taking the fare if they got there first. They could then report the cheat to the taxi controllers, who could then disconnect the offender from the radio network as a punishment. Such a system might seem well suited to an uneasy coalition of mutually suspicious drivers, but it did not provide a solution in Palermo. Gambetta speculates that this may have been because drivers might have been wary of reporting their fellows, in case they picked on one who was under Mafia protection.

Eventually the Palermo drivers agreed on a system in which they used car parks as cab ranks; the dispatcher would identify the rank nearest each caller, and allocate the first in line to pick up the fare. Customers waited longer and their fares were higher, because the meter started running when they called. They were paying the price of endemic mistrust.

Here the Mafia appears to be an impediment to the efficient functioning of a simple commercial service. Yet it may also offer genuine solutions to the difficulties of reaching any kind of a deal between parties certain that each is out to swindle the other. Gambetta quotes a Neapolitan coachman, in 1863, bemoaning his misfortune in buying a dud horse who did not know his way around, missed his footing on slopes, and only wanted to travel the paths he liked. Such a beast was the pre-motor equivalent of a bad used car, the kind that made the second-hand car salesman a clichéd symbol of untrustworthiness. The coachman explained that he would normally benefit from the protection of a *camorrista* (a member of the Camorra, the Neapolitan counterpart of the Sicilian Mafia). 'Last year I wanted to get rid of a blind horse and he helped me to sell it as a good one, for he protected me. Now he is in jail and I was forced to buy this bad horse without him. He was a great gentleman!'[5]

The coachman's lament for his *camorrista* protector affirms that what the *camorrista* provides is a genuine service. His protection is real, sought by the protégé, and missed when it is gone. He is also offering a service that is unreliable but to some degree valuable to those not under his protection, taking a payment from both the buyer and the seller in the horse market for guaranteeing the deal. Had he not been unavoidably absent, he would have seen to it that the coachman did not buy a bad horse. However, he had previously helped the coachman pass a blind horse off as a sound one. This suggests that he sometimes acted as an honest guarantor for both parties, but acted partially in other transactions. When acting honestly he gave buyers and sellers the confidence to buy and sell, thus keeping the wheels of the market turning. When he favoured a protégé at the other party's expense, he reminded all concerned that without his protection they would be cheated.

'The *mafioso* himself has an interest in *regulated injections of distrust* into the market to increase the demand for the product he sells—that is, protection.'[6] In the nineteenth-century horse market the dynamics of mafia activities are already visible. A climate of distrust creates a demand for protection, which the mafiosi emerge to meet. They then develop their market by actions which themselves increase the demand for protection; and protection is on its way to becoming a racket.

Spanish misrule in the eighteenth century cannot be held entirely responsible for the risks of horse-trading in nineteenth-century Naples or the failure of Palermo taxi drivers to establish a radio dispatch system in the late twentieth century. The story is all too plausible, however, as an example of how easily trust is destroyed, and how it may continue to be suppressed indefinitely. Exploitative, unaccountable rulers furthered their own interests by waging war on trust; in the resulting climate of generalized distrust, the demand for protection enabled a new body of exploitative, unaccountable power to develop, which instituted its own methods of dividing and ruling. Many of these—the corruption of officials, the subversion of the rule of law, the undermining of confidence in everyday commerce—are distinctly similar to those of which the Spanish rulers were accused. There may be implications for other regions, especially formerly Soviet ones, which are also marked by the legacy of regimes that stifled the trust between their subjects. A taxi driver who fears that some of his colleagues may be under Mafia protection might well be able to relate to the predicament of a worker in a Soviet-era enterprise who knew that some of his workmates must be secret police informers.

When there is no such thing as trust in society, there are only individuals and their families. If people cannot trust institutions that govern them or associations between them, they fall back on their kin. Trust is abundant within the family but turns abruptly to suspicion at the family's limits. (A mafia is a family that succeeds in turning its dispositions towards insiders and outsiders into power.) Under such conditions it is hard to envisage the rebuilding of community or civil society, even without regular injections of poison into the social atmosphere by a mafia. 'Familistic' values are anti-social, concentrating moral obligations within the family, often at the rest of society's expense. In many familistic societies, the political scientist Francis Fukuyama observes, 'there is a high level of public corruption because public service is often regarded as an opportunity to steal on behalf of the family.'[7]

Fukuyama sees the condition of southern Italy as one instance of a distrustful society produced by the actions of centralizing monarchies upon societies with markedly familistic religious ideologies.[8] Other examples are France, which shares Italy's fervently familistic Catholicism, and China, where Confucianism plays an equivalent role. If a man committed a crime in traditional China, Fukuyama notes, his children were not obliged to report him to the police: 'duties to family trumped duties to the state'.[9] Family values, he argues, underlie structural differences between the economies of China and Japan. He identifies the key to Japan's success as its distinctive traditional readiness to jump the barriers of family ties. Although the Japanese would base their enterprises among kin groups, they were much more open to outsiders than their Chinese equivalents. If a member of a family proved not to be up to the job, Japanese practices were flexible enough to allow an outsider to be co-opted into the household as an honorary family member.

Outside the family, Japanese strangers were exceptionally ready to trust each other even without explicit contractual agreements, and Japanese society developed a strong culture of reciprocal moral obligation. These characteristics remained apparent in the great industrial combines that formed as the nation was reconstructed after 1945, and made Japan one of the world's most prosperous nations. A readiness to trust and to associate across groups could be seen at work in the formation of the distinctive alliances, known as *keiretsu*, comprising clusters of banks and manufacturing enterprises. Within companies, principles of reciprocity assured employees that in return for their commitment, their employer would provide them with a job for life.

In China, however, the family was traditionally on the defensive against exploitative imperial taxation, and the catastrophic political upheavals that followed the end of imperial rule in the twentieth century. Enterprises typically arise and stay within family bounds. Although they can be highly industrious, their reluctance to form external associations limits their ability to innovate. Whereas Japanese enterprises can form new connections and share knowledge in order to develop new products, Chinese ones will merely be good at manufacturing products designed elsewhere. The moral of Fukuyama's account is that imagination and vision depend on trust.

In 1827 Alexis de Tocqueville, a young Frenchman who was to become a renowned commentator on societies overseas, began his travels with a journey to Naples and Sicily. He reported a climate of duplicity and such moral degradation that murder was considered a right.[10] A few years later he visited America, where he encountered a society that had risen above the gloomy Prisoners' Dilemma in which southern Italian society was mired.

The secret of American success was a commitment to self-interest sufficiently developed that its vision extended to the advantages of cooperation. 'Americans enjoy explaining almost every act of their lives on the principle of self-interest properly understood,' de Tocqueville wrote in his celebrated work *Democracy in America*. 'It gives them great pleasure to point out how an enlightened self-love continually leads them to help one another and disposes them freely to give part of their time and wealth for the good of the state.'[11] They understand that the interests of others are entwined with their own.

Americans who behave as de Tocqueville describes will therefore tend to be trustworthy, according to the understanding that trust is an expectation that other parties will incorporate one's interests into their actions. If they are frequently seen to help others and act in others' interests, they will acquire reputations for trustworthiness. If most people are seen to be trustworthy, an atmosphere of generalized trust will prevail. People will tend to assume that anyone they encounter may be trusted, and so will be ready to associate with them, whether to exchange goods, to engage in collaborative projects, or simply to enjoy the pleasures of easy sociability.

The latter is neither trivial nor a diversion. A polite exchange of greetings on a back-country trail reassures each party that the person they suddenly find themselves alone with is a potential source of assistance and not a threat. Similar exchanges in a street between neighbours, regularly repeated, affirm that each acknowledges the other—there is a degree of respect—and that they are benignly disposed towards each other. Remarks made to convey that one is well disposed to the listener rather than to communicate information, such as pleasantries about the weather, are known as phatic comments. They are rewarding in themselves. The child in the opening scene of this discussion will not be the shop's only customer. If the shopkeeper's days are measured out by a series of friendly signals, he can pull down the shutter at closing time and rest contented by these repeated demonstrations that people like him and appreciate their exchanges with him. Regular phatic tokens exchanged between neighbours help to sustain a sense of community, a network of relationships that are mutually supportive and reliable. You know that your neighbour will say hello to you each morning; you form a background assumption that she would equally reliably help you if a situation requiring neighbourly support were to arise.

Neighbourly relations are broadly equal relations, and these were favoured in a nation that had left the old aristocracies behind in Europe. It was the Anglo-Americans' radical equality that impressed itself upon de Tocqueville as the source of their society's vigour. (African-Americans were enslaved in the southern states, and would remain so for more than a generation.) He was particularly impressed by the American enthusiasm for combining to pursue their aims collectively: 'Americans of all ages, all conditions, and all dispositions constantly form associations. They have not only commercial and manufacturing companies, in which all take part, but associations of a thousand other kinds, religious, moral, serious, futile, general or restricted, enormous or diminutive.'[12]

Americans are joiners still: in the 1990s they were only outdone in the density of their associations by the citizens of small countries in northern Europe. When the political scientist Robert Putnam made this comparison, however, he did so in the context of an argument about a long decline of association and community in American life, entitled *Bowling Alone*.[13] Putnam found a retreat from association on all fronts. People were decreasingly likely to join societies in which they met up with other members, to eat meals with other members of their families, or to have friends over to their homes.

Trends like these were indicators of a decline in social capital, the stock of networks that connect people and reciprocities that arise from these connections. The more connections there are between people, the more interactions will take place in which one party will do something for another, who will return the kindness. The more such reciprocal exchanges take place, the more strongly reciprocity will be ingrained as a custom or obligation. People will become more trustworthy; which in turn will encourage further cooperation, to mutual and general benefit.

Putnam points to the gains in efficiency that social systems enjoy as the cycles of reciprocation and trust gain momentum, depicting trust as the lubricant in a reciprocal social engine. 'A society that relies on generalized reciprocity is more efficient than a distrustful society,' he observes, 'for the same reason that money is more efficient than barter. Honesty and trust lubricate the inevitable frictions of life.'[14]

Trust has an intuitively obvious role in cutting what are known as 'transaction costs'. If a newspaper commissioning editor phones a journalist to ask for 1200 words of instant opinion on the topic of the day by lunchtime, and the journalist sets to her keyboard without further ado, the transaction costs for each party are minimal. If the newspaper's corporate policies deem such primitive trust inadequate, however, the newspaper incurs the cost of administering a written contract; the journalist may minimize her time-costs by not reading it, but may later find she has signed up to terms that are not to her advantage. She is faced with the risks and costs that arise when agreements move from a personal to an impersonal basis. Agreements between old acquaintances are cheaper.

Less intuitively obvious, but more profound in their implications, are the associations between social capital and health. In the United States, Ichiro Kawachi and his colleagues found that people living in areas deprived of social capital rate their health as worse than people living in more socially favoured areas; and that people who generally mistrust others are likely to die sooner than people who are generally trustful.[15] Underlying these effects, and social relations in general, is the factor of equality, though in a different aspect to that which struck de Tocqueville so forcefully. American society today is shaped not by the equality of rank its people enjoy in the absence of traditional aristocratic domination, but by the inequalities of income and status that divide them. Societies elsewhere tell the same story: inequality leads to antagonism, illness and death; equality promotes harmony and health.[16]

The best-known findings are those from the 'Whitehall studies' of British civil servants, in which mortality rates (the probability of death during a given period) followed a steady gradient from the top to the bottom of the staff hierarchy. At the bottom of the scale, the lowest-ranking men were three times more likely to die of heart disease than those at the uppermost levels of the civil service grade scale. That was after the possible effects of unhealthy behaviour such as smoking and a poor diet had been taken into account: the effects appeared to arise from the hierarchy itself. Remarkably similar patterns have been observed in other primate species. Baboons sort themselves into dominance hierarchies, and within such a group the subordinate ones are more likely than those who dominate them to show signs of atherosclerosis, the furring of the arteries that leads to heart disease.

This does not in itself identify the hierarchy as the cause of illness. It could be that health determines rank: the fitter primates, including civil servants, rise to the top because they are better able to tackle the challenges that face them, while their sickly fellows sink to the bottom. That is not what is happening, though, as experimental manipulation of rank in monkeys has shown. In one study, Carol Shively and Thomas Clarkson kept female monkeys in four-member groups, assessed their ranks, and then sorted them into new groups, putting dominants with dominants and subordinates with subordinates. The monkeys formed new rank orders, in which some previously dominant individuals became subordinate, and vice versa. Change of rank took its toll on both monkeys that rose and ones that fell, but whereas the effect on the arteries of upwardly mobile monkeys was challenging, the impact on those that were demoted was catastrophic. Dominants that became subordinate were more than ten times as severely affected than subordinates who gained the upper hand, suffering five times more atherosclerosis than their counterparts whose social status did not change.[17]

The damage appears to be caused when a state of emergency becomes permanent. Animals respond to emergencies by readying themselves to fight or to flee. Their bodies are put on a war footing: activities that do not have to be performed constantly are suspended, maximizing the resources available to deal with the immediate situation. If that situation is prolonged, the body suffers from lack of maintenance; it is vulnerable to infection because the immune system is inhibited; and it will suffer the corrosive effects of the hormones that induce stress responses. For a subordinate individual, life in a hierarchy is life in a state of emergency. Each command or challenge reasserts the dominance of others, and confirms the subordinate's lack of choice. Instead of fighting or fleeing, the subordinate simply has to take it. Tellingly, the Whitehall civil servants were less likely to suffer illness to the extent that they felt a sense of control over their work.

They were also healthier if they had a wide circle of friends and saw them regularly. Like the neighbour whose sense of social security is enhanced by exchanging regular greetings with neighbours, the civil servants seem to benefit from a background of sociability. If they thought about it, they would probably assume that there would be ample support from their circle of friends if they needed it, but the point may be that they did not need to think about it.

People also benefit from belonging to clubs or associations, which give them dimensions of status that may compensate for lowly stations at work or in social class. 'Humans belong to multiple hierarchies and tend to value most the one in which they rank highest,' observes the primatologist Robert Sapolsky.[18] A lowly position in the workplace may be 'just a job', but a Sunday role as a referee on a local football pitch is a source of pride—especially since it involves the assertion of dominance over both the players and the spectators. Standing in the community can be more valuable to an individual than standing in a company or society as a whole. The richer the associational life of a community—the greater its social capital—the greater will be its capacity to support the health and well-being of its members.

So too will its ability to promote trust by disseminating information about individuals' trustworthiness. The more forms of association there are, the more opportunities there will be for people to see how an individual behaves when associating with others. In even a small rural community, a person might be able to join a village sports team, dress up in a traditional or modern festival parade, lend a hand to neighbouring farmers informally or within a cooperative, take part in drives to raise funds for the village hall or the church roof, play in a band or sing in a choir, mind a neighbour's children, or spend an evening with other locals in the pub. Although these days many villages may struggle against the socioeconomic tide to achieve such associational vigour, it is probably reasonable to suppose that most villages throughout the world and throughout history have sustained the major dimensions of association—economic, ritual, cultural, domestic, and informally social—in the face of hardship and crisis. It also seems reasonable to suggest that the promotion of trust through reputation has been a universally important function of such activities.

The trust developed through associational life can be divided into two kinds, thick and thin. Thick trust is the sort that grows from personal familiarity, when one has observed over time that another's actions are competently performed and consistently meet accepted moral standards, or that one enjoys favour from another, or that such an individual is considerate towards others in general. Thin trust is based on reputations, norms, and assessments based on signals such as appearance or demeanour. It is what encourages people to think that they will not be misled if they ask a stranger the way to the station, and that they can safely leave their bags by their seats when going to the buffet car on the train. As Putnam points out, thin trust is more useful than the thick variety, for it extends the radius of trust beyond the horizon of first-hand experience. By the same token, it depends heavily on the extent of experience at second hand or further remove. 'As the social fabric of a community becomes more threadbare, however, its effectiveness in transmitting and sustaining reputations declines, and its power to undergird norms of honesty, generalized reciprocity and thin trust is enfeebled.'[19]

Besides enhancing the self-esteem of its participants, a Sunday football league may be an effective means of transmitting reputational information. Players and officials perform in a series of venues around a local league's area, allowing different groups of observers to form perceptions about how fairly they play or officiate. Bad reputations travel farther than good ones, though. Perceptions such as these, based as they are in a system of rival groups, illustrate in minor form a grave problem with social capital.

As noted earlier, tightly knit communities may owe the strength of their ties to their antagonism towards other communities. Divided provinces like Northern Ireland, or Kosovo, do not lack social capital. Their problems arise from having too much of one kind and not enough of another. Networks that knit communities or groups together are known as bonding social capital. Networks that create ties between communities or groups are known as bridging social capital. The metaphor may bring to mind images of the bridge in the Kosovan town of Mitrovica, garrisoned with NATO troops and blockaded by Serbs on the northern bank against the Albanians on the south side, or of Croatian nationalists shelling the old bridge that had linked the Muslim and Croatian halves of Mostar in Bosnia-Herzegovina, until it fell into the Neretva river below.

Radically antagonistic communities pose a particular problem of trust. In an iterated game, the players have a history. Each recalls the other's previous actions and their own. If the game is simple their recollections should be effectively identical, and if the players are computer automata, absolutely identical. They may be at odds but they agree on the facts. In real life, though, players have histories. Each recalls the record of interaction differently; each interprets the data within a different framework of ideas. When the players are communities or ethnic groups, their theoretical frameworks typically identify them as the repositories or defenders of fundamental values: normally moral, often cultural, sometimes biological. Northern Ireland's sectarian division is a relict from the era when much of Europe was torn by conflict between Protestants and Catholics, each side convinced that it was struggling against an opponent that had perverted the word of God. Westerners who see the conflict with political Islam as a clash of civilizations believe the West is worth defending because it is culturally richer and more intellectually advanced than the Islamic world, as well as morally superior. During the era of European colonial expansion, Europeans came to justify conquest with arguments that they were innately as well as culturally superior to the peoples they conquered.

In the fog of group conflict, accounts of the facts are often wildly at odds. People disappear; marauders move at night; there are often more than two sides, numerous factions, and countless liars. Massacres are denied, blamed on others, explained away for decades. Atrocity stories are made up and handed down to successive generations. True and false accounts of unjust treatment and injury are often taken to the heart of group identity. The Serbian attachment to Kosovo, which takes its name from the scene of the Serbs' greatest defeat, by the Ottoman Turks at the battle of Kosovo Polje in 1389, is a notable case in point.

Accounts of injury—in the sense both of narrative and of reckoning tallies—are the basis of a group's assessment of another group's likely behaviour. The system of ideas within which these accounts are shaped and understood, the group's ideology, is the system that articulates and expresses the group's interests. Ideologies give interests global or even cosmic pretensions. They identify the group's interests with those of more general goods, such as civilization, or higher spiritual powers. This, of course, makes them mutually incompatible. God cannot be on both sides; or not in the sense each side imagines, at any rate.

The results are thus more than the differences between players in an iterated Prisoners' Dilemma game who disagree on the tally of cooperative and defecting plays each has made. They are even more than the differences between two corporate rivals who each interpret their records of interaction within the framework of their strategic business interests. Although each party's interests are different, each understands its interests within a framework of economic rationality that the other shares. When a group's understanding of its interests revolves around its belief that its relationship with its god is uniquely close, or that within it resides a uniquely precious spirit, it is in a world of its own.

Within the group, such ontological beliefs—about the essence and fundamental nature of being—are the basis of trust. If members of the group enter into relations with outsiders, whatever trust they develop must be established without this support. In simple exchanges, such as fishermen selling their catch at the harbour or peasants bringing their produce to the market, the requirement for trust may be minimal and ontological considerations should not intrude. But more complex social relations may be unsettled by differences in moral norms, arising from each group's axioms of life, which confound the expectations about another party's behaviour that constitute trust. And deep trust between groups with antagonistic histories may be impossible if the beliefs by which each interprets the world contradict each other. If one's black is the other's white, they are not going to be able to trust each other very far.

Communities kept apart by blocked or smashed bridges are not typical, however. A number of striking examples have appeared out of the political turbulence of recent years, but they are against the broader currents of history, those of mass migration. Modern cities contain multitudes of people and multiple cultures. Even modest towns have minorities. It is not just a question of whether all these people can stand each other, or even whether they can just get along, but of whether they can really live together. Can they become a community, or must they remain foreign to each other despite living in each other's presence?

Research by Robert Putnam and his colleagues suggests that building communities out of diversity will be an uphill struggle. Based on a large survey of districts in towns and cities across the United States, and drawing upon studies from other parts of the world, Putnam's account depicts a decidedly negative relationship between diversity and community.[20] High ethnic diversity is associated with low social trust in reports from the United States, Australia, Sweden, Canada, and Britain. In Prisoners' Dilemma and Ultimatum Games, 'from Uganda to the United States', players who are different from each other are more likely to defect.

Within the United States, diversity has a depressing effect on collective life. In areas of greater ethnic diversity, people trust each other less. About 30 per cent of San Franciscans, who live in one of the most diverse localities in the world, say they trust their neighbours a lot. In a rural South Dakota county where 'celebrating "diversity" means inviting a few Norwegians to the annual Swedish picnic', the figure is around 80 per cent. People also have less confidence in local political structures. They are less likely to take part in community projects, to do voluntary work, or to give to charity. They have fewer friends and watch more television.

These really are effects of diversity, Putnam emphasizes, visible even when neighbourhoods with equal levels of poverty or crime are compared. When people of different groups live amongst each other, the dearth of capital bridging the groups does not intensify the bonding capital within them. 'Diversity seems to trigger *not* in-group/out-group division,' says Putnam, 'but anomie or social isolation.' People in ethnically diverse settings withdraw from each other. Putnam compares it to a turtle pulling its head back into its shell.

The turtle's reaction is an automatic reflex. Putnam's metaphor seems to imply that humans are likewise impelled by an instinctive aversion towards others whom they regard as not their own kind. He does cite a study from Britain, however, which challenges this perception.[21] Natalia Letki notes that the most popular explanation for the inferior quality of inter-racial exchanges, which are found to be less honest and less reciprocal than ones within ethnic boundaries, is that individuals living in heterogeneous neighbourhoods interact with each other less because they prefer people they regard as similar to themselves. Trust and reciprocity suffer in consequence. She finds that diversity does have a negative influence on just one element of social capital: people's attitudes towards their neighbours, including their trust in them—yet they have just as much to do with each other as they would if they all belonged to the same ethnic group. Their actions and their words don't match. Perhaps it is not diversity that is the obstacle to trust, but prejudice.

Letki also finds that trust and other neighbourly dispositions are the only aspect of social capital that diversity seems to impair. Other apparent effects are dispelled when proper account is taken of the material circumstances of diverse neighbourhoods. Economic deprivation, not ethnic diversity, is what erodes social cohesion.

Trust is also a legacy of history. Where people from diverse backgrounds do become involved in institutions, contradictory histories may hamper their efforts to work together. If a minority group has a history that leads it to mistrust the police, members of that group serving on juries may be sceptical about police evidence, and find themselves at odds with fellow jurors whose trust in the criminal justice system is resistant to question. In some settings this polarization is stark. When the black American football player O. J. Simpson was acquitted of double murder in 1995, African-Americans appeared much readier than whites to consider the verdict sound. This difference in opinion was widely taken as an indicator of the depth of African-Americans' alienation from a system that is supposed to ensure justice for all.

Elsewhere the gap in perceptions seems to have narrowed. In Britain a generation ago, distrust of the police was largely the preserve of ethnic and political minorities, along with some working-class communities. Middle England dismissed them as malcontents or ne'er-do-wells. Public perceptions may well have been forced to shift, however, by official recognition that three groups of prisoners (known as the Birmingham Six, the Guildford Four, and the Maguire Seven) were innocent of the IRA terrorist crimes for which they had served up to sixteen years in prison. More recently, minority complaints were endorsed by an official inquiry into the investigation of the murder of the black teenager Stephen Lawrence, which concluded that 'institutional racism' permeated London's Metropolitan Police and other police services.[22]

Nowadays it seems to be more broadly accepted in Britain that the probity of the police should not be taken for granted. Although miscarriages or failures of justice must have played their part in this shift of opinion, it is part of a much broader trend. Even when people live in stable, democratic countries where they grow steadily more prosperous, their attitudes towards institutions are typically querulous and their trust in politics hollow.

6

THE GOODWILL OF THE PEOPLE

One morning in March 1985, a group of people assembled outside a small house in the capital city of a small European country, and began planting trees. This was a coup for the musicians' organization that occupied the house, and a bravura display of bridging social capital, for the party included the novelist Kurt Vonnegut Jr and an entourage of diplomats from the American embassy. Vonnegut planted a tree, and a couple of months later John Updike did likewise. But this being communist Czechoslovakia, the government cut down the trees, and several of the organization's leading figures were jailed the following year.[1]

Three years after that, the communist regime itself collapsed; the musicians' organization, originally known as the Jazz Section, is still going strong. Its history is a model of how states should not conduct relations with civil society, and is part of a larger history that has aroused much of today's concern about trust in political institutions.

The Jazz Section was a product of tensions between civil culture and Leninist power that the regime spent two decades failing to resolve. Under the Soviet doctrine imposed on Czechoslovakia after the Second World War, associations were supposed to be channels through which the Communist Party could organize, discipline, and shape the people. They were part of the state apparatus, whose purpose was to 'build socialism'. The difficulties experienced in this project led, however, to a brief period in which many of the tenets of the Leninist system were thrown open to question, offering the vision of 'socialism with a human face'. After five months of this 'Prague Spring', Soviet-led forces crushed the reform movement in August 1968.

Attempting to restore the *status quo ante*, the new satellite regime decided that Czechoslovak social organizations had become too independent-minded to be brought to heel, so it disbanded them and set up new ones. In 1971, a group of jazz enthusiasts applied to the Ministry of the Interior for permission to form a jazz musicians' union. The ministry officials turned them down, but recommended that they join the new Czech Musicians' Union. Two crucial lapses gave them a degree of associational freedom markedly greater than they were supposed to enjoy. First, the Musicians' Union misinterpreted the Ministry, which had intended that the jazz musicians join as individuals, and allowed them to set up their own Jazz Section. Second, the Ministry was not given the power which it normally exercised over social organizations to appoint the Section's chair.[2] The authorities did, however, impose a limit on the number of people who could enter the association, as if it were a bus.

The new Jazz Section set to work exploiting the limited trust placed in social organizations by the state. Since such associations' interests were assumed not to be of wider interest, they were permitted to publish outside the normal system of censorship. The Jazz Section's publications extended beyond jazz, however, into art theory, literature, and other areas, creating a space for the exchange of non-conformist ideas. They were phenomenally efficient as bridging social capital, read by perhaps ten times as many people as the Section had members—who themselves numbered more than double the official maximum capacity of 3,000. After the Section's leaders were arrested on charges of illegal publication and commercial activity, Party members were said to have sold each other confiscated copies of its edition of Bohumil Hrabal's novel *I Served the King of England*.[3] In its perverse way, that was also a kind of bridging social capital, albeit a bridge too far.

Against the face of the repressive bureaucracy of the 'normalization' regime, the Jazz Section offered a glimpse of the creative power of civil society. Lessons like these were not lost on dissident thinkers such as Václav Havel, the writer and dramatist who became president after the collapse of the Czechoslovak communist regime. The seeds of civil society's prominence as a theme in public discourse were sown in Soviet-controlled central Europe. After the bloc disintegrated, conditions in the emerging democracies also threw the spotlight on the allied subject of trust. These were global issues, as nations on several continents made the transition to market democracy. Meanwhile, within the established democracies of North America and Western Europe, anxieties grew over declines in the trust enjoyed by politicians and their institutions.

Moscow's regime and its satellites left a legacy of depreciated industrial capital and minimized social capital. The residues of each have proved toxic. Soviet doctrine not only denied the value of social capital, but did its best to straitjacket any associations that threatened to produce it. Culture and sport were institutionalized by the state, while the opportunities for informal socializing were limited by ramshackle economies as well as by the official preference for disciplined recreation. As in southern Italy, a regime at odds with society forced people to retreat into their families.

It also institutionalized dishonesty, forcing people to adhere to official phraseology in public and to only say what they really thought in private. When a clerical worker asked a departing foreigner whether he might come back to 'help build socialism', was she being sarcastic, seeking titbits of information about foreigners to gain credit with the secret police, using a politically correct phrase to avoid being reported to the secret police by her colleagues, or echoing the rhetoric because she genuinely believed in it? By the 1980s in Poland it would almost certainly have been the first, but in East Germany it might well have been the last, right up to the end. Wherever, and at whatever stage in the bloc's history, one could never be entirely sure. For those without the freedom to leave for the West, the consequences of misreading political remarks could be drastic. Trust in the public sphere was prohibitively risky. Everybody was dissimulating to safeguard their own interests except for the true believers, who were the least trustworthy of all.

The party-states have also left a legacy of anti-social capital, in the form of networks of suspicion about who collaborated with the old regimes. These are extensively politicized, being used by various factions to discredit many of the former heroes of the dissident era. Access to their material base, the vast records left by the security services which monitored the people of the People's Republics, is bitterly contested. The files are the records of thousands of iterated Prisoners' Dilemmas that faced people living in police states. Some factions agree with Tadeusz Mazowiecki, Poland's first non-communist prime minister of modern times, who called in his inaugural speech for 'a thick line between us and the past'.[4] Others insist that communism will continue to corrupt states and societies from beyond the grave, until there is a reckoning of accounts.

Among Polish conservatives and nationalists this shades into conspiracy theories which claim that Poland is in fact still run by the hidden hands of the security services, in collusion with formerly communist politicians who have adapted successfully to free elections. The communists distrusted the people; the implacable anti-communists distrust the organs of power with equal fervour. Nationalists also feel that the people have let the nation down, not just by voting for 'post-communists' but also by a more general drift from traditional authority. They do not trust the people to uphold the nation's interests, which they conceive in spiritual as much as material terms.

It should be added, though, that Poles don't have to be nationalists to distrust each other. Migrant Polish workers in London, some of the hundreds of thousands who came to Britain after Poland joined the European Union in 2004, complain about each other rather than about the locals.[5] *Polak polakowi wilkiem*, a Pole is a wolf to Poles, they say. This bitterness reflects the Hobbesian conditions in the lower reaches of a free labour market, and migrants' vulnerability to exploiters who speak their language. It also illustrates the semi-detachment of national identity from social trust. People blame their compatriots for failing to live up to national ideals: the nation, like the Queen of England, is above criticism.

Behind the nationalists' jaundiced view of democracy can be glimpsed the shades of the military moralists who ruled many European countries during the twentieth century, certain that it was better to make the people adhere to the doctrines of church and nationhood than to trust them to make what they wanted of society. And their scepticism is shared by a large fraction of Polish society. In an opinion poll taken at one point in 2005, more than half the respondents agreed that undemocratic government could be better than the democratic kind. Thirty to 40 per cent consistently disagree that democracy is superior to all other forms of government.[6] Similar percentages have declined to endorse democracy in other formerly communist states, including Russia.[7] Whether they doubt their fellow citizens' public spirit, or the ability to channel public spirit through democratic channels to good effect, they do not seem to trust their fellows as citizens very much.

Would they be more inclined to trust their new political institutions if they trusted each other more? Among social scientists, two contesting views have arisen. On one hand, 'institutional' theories take a relatively straightforward view of political trust. They regard it as a rational response to institutional performance. If institutions are seen to work efficiently and honestly, people will trust them, and if they aren't, people won't. As trust is a response rather than an influence, it will not make a great deal of difference to political systems whether people trust them. On the other hand, 'cultural' theories see trust as a source of support for democracies, encouraging people to embrace democratic values and to become politically involved. They also believe that trust tends to spread. People come to trust each other through spending time with each other informally, which encourages them to join together in associations. The more their personal experience assures them of people's goodwill and ability to cooperate at the grass roots of civic association, the more inclined they are to trust higher and more powerful political associations. This is de Tocqueville's vision of democracy in America, brought to contemporary America's notice by Robert Putnam, and tested by researchers in varyingly democratic countries around the world.[8]

How the relationship between social and political trust looks depends on where you stand. In the United States, the scene is backlit by the civic ideals of the nation's formative years. Putnam's vision of the decline of associational life in the United States is an elegy for the passing of the great civic generation, born between the end of the First World War and the early years of the Cold War, which stands out from his data as uniquely blessed with community and public spirit. The civic generation joined associations enthusiastically, lent its hands to good works gladly, voted optimistically, and in youth looked forward to a life of service rather than a career. The decline or extinguishing of those citizens' energies offers an explanation for the simultaneous decline, over several decades, of trust in American political institutions.[9] The good citizens of the middle twentieth century affirmed that an ideal America was possible. But its ideals may not translate, as Martin Amis may have meant to suggest when he remarked that in *Star Trek*, the *Enterprise*'s officers boldly go 'to pester alien life forms with the American Constitution'.[10]

In Istanbul or Moscow, by contrast, activism is likely to go with alienation. People are more likely to engage in political action if they distrust institutions, believe that public life is steeped in corruption, and see political parties as machines for furthering oligarchs' interests.[11] And not so long ago, this is how the West was won. Now that protest in liberal democracies is to a large extent a publicity exercise, it is easy to forget that many constitutional rights, such as votes for women and civil rights for black Americans, were established in the wake of direct action by groups convinced that trusting institutions would get them nowhere.

Democracy itself was brought to many countries through direct action by popular movements, mass associations that can spring up overnight and mobilize millions for collective action. They often seem to create social capital out of thin air. On the civically icy terrain of communist Poland at the beginning of the 1980s, the Solidarity trade union movement brought together nearly ten million people in a self-organized network that utterly eclipsed the official trade union federation. Among them were a million of the communist party's three million members. In a movement that included pockets of striking workers who had to be dissuaded from singing about hanging commies from lamp posts, this was a remarkable achievement in bridging social capital.

The movement as a whole raises intriguing questions about trust. People suddenly started to act as though they all trusted each other. They embarked on ventures that entailed great risk, both locally and nationally. Everybody was listening out for the rumble of tank engines from the other side of the Soviet border. Eventually the Polish government suppressed the movement by declaring a 'state of war', amounting to an internal military occupation.

Despite the risk that individuals could defect from the movement as the collective risks increased, Solidarity lived up to its name. There were certainly a number of possible sources of trust to make up for the denuded social landscape: shared Catholic faith and the Church infrastructure, a vivid and passionate shared sense of national identity, links between workers and dissident intellectuals, and a powerful working-class consciousness based not just on the experience of heavy industrial labour, but also on what the regime had been telling the workers about themselves for several decades. Above all, individuals could be confident that most people they encountered were likely to share their opinion of the government, and therefore to see their interests similarly.

Revolutionary moments like this one are not part of the normal order of things, though, and reciprocity may not work by the normal rules. These are emergencies, and in emergencies people voluntarily decide to defer the reckoning. They cooperate first and worry about defectors later. Rather than trusting each other, they may decide to run the risks of co-operation without considering whether they trust each other. Collective solidarity demands this as a duty, and then builds trust upon it.

Once the moment is passed, the people's emergency powers of association lapse. Even if they achieve democracy, their associational life may not continue to strengthen it. In South Korea, the more that people are involved in associations, the less inclined they are to trust the parliament and the parties that replaced a military regime after mass democracy protests in 1987. On the other hand, associations in South Korea do seem to have a bridging effect: the more people are involved in them, the more likely they are to trust their fellow citizens. Ji-Young Kim, who published these findings, comments that association in South Korean society works upon social trust as Robert Putnam would expect, but not on political trust.[12] Other newish, or questionable, democracies also seem not to work as Putnam and de Tocqueville would hope.

The Putnamesque and Tocquevillian view of the relationship between society and state is also qualified by research into one of the world's oldest democracies. Putnam draws attention to the influence of the grass roots upon the institutions of government: associations among the people develop trust, which wells up into the structures of power. Markus Freitag uses data from Switzerland to argue that the state can be a producer of trust as well as a recipient of it. He finds that the

Swiss use of direct democracy, allowing citizens as well as their elected representatives to vote on decisions, promotes 'a lively associational life'. Switzerland is divided into cantons, which vary somewhat in their political arrangements: the more that the citizens of a canton are able to participate in political life through submitting initiatives or voting in referendums, the livelier their associational life is likely to be. Opening up access to the structures of power, Freitag suggests, may be a way to break out of 'the vicious circle of distrust, disengagement and weak democracy'.[13] This may be so, but it is worth recalling that five hundred years of democracy did not provide access for women to the federal parliament, either as members or voters, until 1971. Switzerland was the last country in Europe to allow women the vote, apart from the tiny principality of Liechtenstein.

Putnam's selective attention to society as the source of trust may be seen as part of an ambivalence that pervades much American thinking about the state. When the English barons rebelled in 1215 against King John's oppressive rule, forcing him into the agreement set down in Magna Carta, they enacted a constitutional limit on the ruler's power. Subsequent statecraft refined the drafting of constitutions, but they continued to be based on the premise that rulers would abuse power if they were not contractually restrained from doing so. In the eighteenth century, liberal thinkers such as David Hume, John Locke, and Adam Smith were keenly aware of that danger. 'Political writers have established it as a maxim,' Hume noted, 'that, in contriving any system of government, and fixing the several checks and controuls of the constitution, every man ought to be supposed a *knave*, and to have no other end, in all his actions, than private interest.'[14]

At the same time the liberals accepted the necessity of the state, not least, as Thomas Hobbes had argued, to protect the people from each other. It had to watch over the citizens, but it had itself to be watched. The result in North America was, in Russell Hardin's words, that the 'United States was created on a principle or theory of distrust in government'.[15] Hardin emphasizes that the principles at stake were economic as well as political: James Madison, the 'Father of the Constitution', wanted the federal government to curb the state governments' interference with commerce. Today the federal government is itself widely seen as a self-serving insult to the principles of the market. It is not uncommon among Americans to find ardent patriotism combined with a deep distrust of the government apparatus that makes a nation of their states.

The principles of balance by which modern states are designed apply as much to mistrust as to any other quality. Although mistrust of agents in positions of power is fundamental to the liberal state, excessive mistrust may have pathological effects. Mistrust of the federal government is often said, for instance, to encourage people to keep guns. A belief in self-reliance, that one should be ready to defend oneself and one's own instead of relying upon the police or other state forces, is acidified by a strain of distrust in the state and its forces, souring at the extremes into a suspicion that those very agencies may turn out to be the enemy for whom one must be prepared.

Though the reasons are unexplored, there is some evidence indicating that Americans who lack faith in the federal government are more likely to own guns than those with greater confidence.[16] It is reasonable to speculate that these mistrustful gun owners increase the prevalence of firearms indirectly as well as directly, inducing others to regard gun ownership as normal, and perhaps also encouraging the perception that it is necessary to arm oneself because so many people around one are armed. It also seems reasonable to suggest that mistrust of the federal state thereby contributes to the toll of casualties caused by guns in the United States.

The toll will also be influenced by the manner in which guns are acquired, kept, and used. At a time when nearly half of Americans and over a quarter of Swiss possessed a gun, firearm homicide rates in Switzerland were less than a tenth of those in the United States—a figure more remarkable for the fact that many Swiss kept assault rifles in their homes.[17] These weapons belonged to the federal government, which entrusts them to men of military age until the latter have completed their service obligations. Switzerland's doctrine of national defence is based upon the organization and arming of the populace into a 'well regulated Militia' of the kind to which the US Constitution refers in asserting the right to keep and bear arms. The country, formally termed a confederation, is divided into several language groups as well as its cantons, but the militia system speaks of a fundamentally trusting relationship between federal state and citizenry. In recent years, however, militia numbers have been reduced; and shooting outbursts of the kind known in America as 'spree killings' have aroused public concern about the extent to which the state's trust in its armed citizens is misplaced.[18]

Distrust is the modern political condition. Trust enjoys much of its profile as a topic of public discourse to hand-wringing about the lack of popular trust in politicians. Vows to rebuild trust have become a platitude for politicians taking office, such as Gordon Brown, who took over from Tony Blair as Britain's prime minister in 2007.[19]

Brown did not have to rebuild trust eroded by poor competence. The Labour Party's victory in three general elections was based upon a steady and confident economic performance for which he, as Chancellor (finance minister) throughout the whole period, could take considerable credit. He may have wished to send a coded message to the large number of voters who believed that his predecessor had lied to them about the reasons for invading Iraq, but his freedom of manoeuvre in that regard was limited by his own record of support for the attack. His declaration took the generalized form that fits pretty much any prosperous market democracy, populated by querulous citizens who are disinclined to give their politicians credit for historically unprecedented levels of prosperity and security. The more comfortable people grow materially, and the more remote grow the prospects of war with neighbouring countries that haunted most of Europe until the day before yesterday, the less likely they are to trust the politicians who preside over this state of affairs.

Britain also illustrates how the public may distrust those whom they elect while trusting much of the state that polices them. British citizens do not in general fear the knock on the door before dawn. Four-fifths of them trust the police.[20] They have acquiesced in the installation of surveillance cameras—except the ones that photograph speeding vehicles—to the extent that Britain is now said to have more of these cameras per citizen than any other country. At present, according to a poll conducted in 2006, most people appear not to feel that the state holds too much information about them. Only a small minority opposed proposals to introduce identity cards when they were first announced, and even at a point when support had dropped from nearly 80 per cent to 50, only one person in six opposed them on principle. Support declined as the government shifted its arguments for the scheme away from countering terrorism, and as stories appeared about how much citizens might have to pay for their cards. Although the public appears comfortable with the status quo, its trust in the state is not indefinitely elastic. There is widespread unease about the prospect that the state will acquire a great deal more information about individuals as it enters details from the cards into a national database. Neither politicians nor civil servants are trusted to keep such details confidential.[21] This was the case even before Her Majesty's Revenue and Customs lost discs containing personal details of 25 million people—nearly half the population—and their bank accounts.[22]

Even if people anticipate that individual civil servants might leak confidential information, either through ineptitude or venality, this is unlikely to make them perceive that the system itself is against their interests. They may be perturbed in the abstract by the possibility that information about them held on a database might be wilfully misused, but it takes a speeding ticket to really raise their hackles. People who object to speed cameras are fond of claiming that the devices serve the interests of the state, as 'government cash machines', rather than public safety. The popularity of this argument, shifting the charge of cheating from lawbreakers to state, basks in the strength of the presumption that government is self-interested to the core.

When they get out of their cars, people's perception of their interests changes. They feel vulnerable, and they welcome surveillance cameras as protection in public space. Surveillance may be seen as a mechanism by which the state can promote social trust. As the political scientist Margaret Levi notes, the first requirement for the creation of interpersonal trust by the state is the ability to observe the behaviour of those subject to the government's laws and regulations, so that lawbreakers can be sanctioned.[23] If people feel that others will be deterred from misbehaving in public by the presence of cameras, they may be more inclined to trust the strangers around them. 'Smile—You're on CCTV', large signs on police vans advise, promoting civility through surveillance.

Intriguingly, people appear to be as firmly against the use of surveillance microphones, used to monitor conversations in public places, as they are in favour of surveillance cameras.[24] They are treating the state as if it were a person. Under the conventions of public space we cannot be expected to avoid looking at people, at least in a general fashion, but listening to other people's conversations is considered intrusive. The difference is similar to that between surveillance and spying.

This tendency to personalize public agencies is a detail of a much more pervasive phenomenon in contemporary politics. It arises in part from the conversion of political parties from movements into enterprises that compete like businesses in consumer markets. Parties traditionally arose as expressions of class or ethnic interests, and sought to involve as many as possible of their 'natural supporters' as activists or members. In many places they became part of the fabric of society, taking the initiative in cultural as well as political and economic activity. One may see the residual signs of these vibrant political associations in Italian towns, where party offices adorned with emblems still occupy prominent places on the main thoroughfares.

In the United States, the bond between voters and the two great parties, Democratic and Republican, is typically enduring. But the relationship appears to be one that forms between individuals and parties, rather than growing out of communities. People decide in early adulthood which they prefer, and their preferences may well be fixed for life. Shifts in the balance of support for the two parties thus reflect changes in the demographic balance between cohorts of voters whose political loyalties were formed by different historical contexts.[25] This implies that failures of political trust, such as the Watergate scandal that forced President Nixon from office in 1974, or the widespread belief that the Bush administration intentionally misled the public when preparing the ground for the invasion of Iraq in 2003, could still be ghost issues in elections half a century later.[26]

Elsewhere, however, many people traditionally felt that their political affiliation was no more a matter of choice than their religious denomination. 'I've voted Labour all my life,' many working-class Britons would say, making a statement about the immutable nature of their political identity. It was partly a matter of perceived class interests, the Labour Party being seen as the political organ of the working classes, and partly of solidarity: to vote Conservative or Liberal would be a form of defection. To their activists the trade unions and the Labour Party were 'this great movement of ours', a phrase that became such a ritual incantation at conferences that it was waggishly contracted to 'thigmoo'. The movement commanded loyalty not just because it pursued its class interests, but because in doing so it was expected to create a good society, harmonious, prosperous, and just. According to Harold Wilson, whose tenure as Labour Party leader and prime minister is not remembered for its idealism, 'This Party is a moral crusade or it is nothing'.[27]

That great movement of theirs was not a conspicuously trusting one. It was based on a practice of policing and sanctions against defectors. If workers tried to defect from a strike, pickets would stand in their way and try to stop them, sometimes forcibly. Anybody who did cross a picket line might be ostracized by other workers and the community in general. Some former miners who worked during the great strike of 1984–5 still are. Likewise the trade unions of Harold Wilson's day were ready to help get Labour into power, confident that their industrial muscle could halt any government actions that they took as defection. British labour solidarity was based on instilling norms, from the day the apprentice arrived on the factory floor and learned to respect the tea break, rather than on any readiness to suspend reckoning.

The old movement petered out as the industries in which it was based went into decline. 'I've voted Labour all my life' had a different significance coming from somebody who had been a car worker or a miner all his life than it does from somebody who has worked in a factory, a supermarket, a minicab firm, and an old people's home. Nowadays it just describes a series of choices made once every few years. Major political parties across the spectrum have decided that relying on a sectional base is not a viable strategy. They seek to appeal to voters across classes, as individuals. Accordingly they employ the techniques of persuasion used by commercial enterprises to market products to consumers. Choosing between parties has become more like choosing between banks: there are few differences in beliefs about economic policy, but the detailed offers and the organizations' performances vary. Voters must trust in a government's technical competence and basic professional integrity, but in Britain at least they no longer need to worry whether a party can be trusted to pursue a moral crusade.

With the atrophy of political movements, political allegiance is resolving into two atomized terms: the individual voter and the individual who stands to become the nation's leader. It has become a commonplace that parliamentary elections have become more 'presidential', revolving more and more around the television images of those who stand to become prime minister. Politicians are deemed by party machines to be good prospects if they seem like good company, appear able to express their emotions or to attend to others expressing theirs, and look at ease with themselves. Voters are encouraged to consider the candidates as if they were choosing a fishing companion or a dinner date.

They are thus tuning in to a notorious array of signals, well known through gossip, popular culture, and, usually, personal experience to be unreliable indicators of a relationship's prospects. Yet costly signals are persuasive signals, and these ones are embedded in some of the costliest displays ever staged. In the United States, candidates' chances of success are measured by the sizes of their campaign funds. An election campaign is a theatrical exercise in seduction, misdirecting personal fantasies into a civic context. People say they don't trust politicians, yet they appear to suspend their mistrust while they are being courted, to enjoy the attention, and to take the opportunity to let their fantasies play a little. Often they are disappointed, but next time around they are willing to go through the process again.

Perhaps that isn't surprising, since this is also now the pattern for so many of their intimate relationships. Trust may be beyond them, but people can still live in hope.

7

LEAVING THE DOOR UNLOCKED

'GENERALLY SPEAKING, would you say that most people can be trusted or that you can't be too careful in dealing with people?' Generally speaking, this question is the instrument that social scientists rely upon to measure what they call generalized or social trust. While other questions come and go from the US National Opinion Research Center's General Social Survey, the trust question is a constant. But what does it mean to the people who answer it? Would they say that most people could be trusted to give them directions, or to mind their bags, or with their cash machine identification numbers? Are they thinking of most people they know, most people they pass in the street, or most people in the world?

Curiously, one investigation found that they might actually be thinking of themselves.[1] Researchers assembled pairs of Harvard undergraduates to play a version of the Trust Game, described in Chapter 2, in which one player is invited to send money to a second, whereupon an extra sum is added and the recipient is given the opportunity to return money to the sender.[2] Each subject was also told that an experimenter would drop an envelope, addressed to herself or himself and containing ten dollars, in a public place. They were asked to put a price on the drop, their trust that strangers would return the envelope to them being measured by how much they would pay for the exercise.

The results of the experiments were compared with the results of questionnaires designed to indicate how trustful and trustworthy the subjects were. 'Standard attitudinal survey questions about *trust*,' the researchers reported, 'predict *trustworthy* behaviour...much better than they predict trusting behaviour.' When in the role of sender, subjects who agreed that 'most people can be trusted' sent a statistically negligible 22 cents more than those who opined that 'you can't be too careful in dealing with people'. As recipients, however, the more strongly they indicated that they trusted people, the more money they returned to other players. 'To determine whether someone is trustworthy,' the researchers concluded, 'ask him if he trusts others.' The uplifting moral undertone of this message was offset, however, by the finding that people sent back less money when paired with players of different race or nationality. In twelve instances the recipient sent back nothing; the other player was of a different race in all but one of them.

The trust question itself emerged from the study in a peculiar position. It had been identified as an effective instrument for determining trustworthiness, but was only weakly associated with the trust about which it ostensibly inquired. And the more widely it is asked, the more we may wonder about what people mean by their answers. In one survey, 65 per cent of Norwegians answered it affirmatively, compared to only 3 per cent of Turks. (The figure for Britain was 30 per cent, equal with that for South Korea and six points lower than that for the USA.)[3] We may reasonably infer that the Norwegians are slightly more trusting than the neighbouring and culturally similar Danes, who scored 58 per cent, and it certainly appears that Peru and Turkey, each scoring 5 per cent, are deeply distrustful societies, but how confident can we be that Norwegians, Peruvians, and Turks are answering the question according to the same terms? Frames of reference may vary from place to place, as in the question of happiness: in some places, people may consider themselves happy to have enough to eat and not be caught up in a war; in others, they may not be happy with anything short of heaven on earth.

Russell Hardin thinks there are also a couple of odd things about the phenomenon of generalized trust that this 'badly framed, vague' question is supposed to detect. Firstly, it sounds more like optimism than trust. Second, saying that one trusts most people most of the time is an 'elliptical' claim. Common sense tells us that it might go as far as minding bags, but certainly not as far as cash machine identification numbers. What seems on the face of it to be a rather strong declaration of general confidence in humanity is really just a mild expression of preparedness to take minor risks in everyday dealings. Hardin is sceptical about generalized trust because he insists that a trust relationship has three parts, A trusts B to do X, and is not meaningful if X could be anything. At best, generalized trust is 'the stance... of the child who has grown up in a benign environment in which virtually everyone has always been trustworthy'. Its value is the value of such a childhood, which encourages people to enter into relationships with others because they have little sense of being at risk from them.[4]

Hardin's brisk dissection of generalized trust is a typically salutary example of the scrutiny to which he and his colleagues have subjected trust in the books they have produced for the Russell Sage Foundation in New York. Trust is a feel-good sensation and an idea that is hard to pin down. The idea of generalized or social trust demands particularly close attention. It is all too easy to see Hardin's point, but equally easy to see that generalized trust describes something that is at the heart of a good society, and is not to be lightly dismissed.

Even if it does consist of minor risks in everyday dealings, those risks and dealings add up. They set the tone for interactions, promoting civility and reducing tensions. They regulate the social climate, making it conducive to associations and relationships of different kinds. And if generalized trust is prevalent, it is not naïve. Where most people are inclined to think that most people can be trusted, they are likely to be right. (If the Trust Game results described earlier have identified a general phenomenon, they will be right because people who say others can be trusted tend to be trustworthy themselves; but they will not profit from it because they don't act as though others can be trusted!)

Generalized trust has also been described as a 'standing decision' to give most people the benefit of the doubt.[5] Yet this is not trust in either the formal, three-part sense or the intuitive one, in which we recognize trust as an involuntary feeling like love. A standing decision to give the benefit of the doubt is more a matter of principle, or duty. It is often encouraged by religious convictions, and should not seem especially paradoxical to those raised in the Christian tradition that enjoins them to 'love your neighbour as yourself'. Others may take a secular route to their standing decision, if they have a basically optimistic view of human nature and understand the power of reciprocity. Social or political idealism may also encourage them to assume that others will treat them as they would wish to be treated. The decision is to make oneself vulnerable in order to build trusting relationships, which will reduce vulnerability by increasing the probability of cooperation. With reciprocity, if you care for your neighbours, you should come to care about them, and if you act as though you trust others, you should come to trust them.

The American political scientist Eric Uslaner describes this as the moral foundation of trust.[6] He distinguishes between placing faith in strangers, which he calls moralistic trust, and having confidence in persons one knows, which he calls strategic trust. According to Uslaner, the formula for generalized trust comprises not three terms, A trusts B to do X, but just one: A trusts. Generalized trust is based on moralistic trust, the standing decision to give the benefit of the doubt, but is influenced by experience. It is 'the perception that *most* people are part of your moral community'.[7] We conceive our moral community, Uslaner argues, upon the premise that although its members may adhere to different religions or secular ideologies, they hold fundamental values in common. This would seem to require an optimistic view of human culture as well as of human nature.

Indeed, according to Uslaner, generalized trusters are generally optimistic. They are happier than mistrustful people, confident that they are in control of their destinies, welcome dealings with strangers as opportunities rather than threats, and are tolerant of people who are different from themselves. By contrast particularized trusters—ones who trust only those they see as their own kind—are self-centred, have difficulties in establishing personal relationships, feel threatened by the outside world, tend towards paranoia and lean towards authoritarianism.

This sounds like a psychologically coloured portrayal of the divide between liberals who welcome globalization and the conservative nationalists who recoil from it. Uslaner places greater emphasis, however, on a dimension less salient in contemporary political debates. Democracy and trust have an uneasy relationship: democracy makes trust possible, but democratic societies are not necessarily very trusting ones, and distrust of power is inherent in democracy. The most important social determinant of trust, Uslaner finds, is economic equality. Unequal societies are mistrustful societies; equal societies are trusting ones. Equality is a cause of trust, not just an association: economic equality leads to trust, but trust has no influence on equality.[8]

Uslaner and his colleague Bo Rothstein note the curious absence of inequality from the academic literature on social capital and social trust.[9] It is much more pervasively absent than that. Despite the impressive body of evidence that inequality damages individual and social health, noted in Chapter 5, public discourse and popular concerns seem to look straight through inequality without seeing it. Even where inequality is remarked upon, it is not recognized as a problem in itself. It may be used as an index of deprivation by which the neediest in society may be identified and provided with state assistance. In contrast to the systems of universal welfare provision for which the Nordic states of Europe are still famous, this is neither a trusting nor a trust-promoting strategy. As Rothstein and Uslaner observe, means-testing for benefits casts aspersions on recipients. It is a strategy preferred in societies where the haves distrust the have-nots, tend to regard them as solely responsible for their predicaments, and seek to identify the deserving among them. Universal welfare is based on a presumption of trustworthiness, or at least a decision to give the benefit of the doubt. The British state pays child benefit for all children, without attempting to satisfy itself that parents will not spend it on cigarettes or debauchery. On the other hand, local authorities only provide free school meals for poorer children, providing a means by which their classmates can identify them as a separate and stigmatized group.

Rothstein and Uslaner draw social-democratic implications out of their work, asking whether low levels of trust and social capital might be caused by a lack of government action to reduce inequality. You don't have to be keen on the state to recognize the central importance of equality, though. The science writer Matt Ridley proclaims its relationship to trust as the culmination of an argument against the power of the state, for which he has a peculiar horror: 'We must encourage social and material exchange between equals for that is the raw material of trust, and trust is the foundation of virtue.'[10]

Trustful societies have a distinctive profile, according to Jan Delhey and Kenneth Newton, who are based in Germany and England respectively. Highly trusting countries are ethnically homogeneous, are well governed, have Protestant religious traditions, and enjoy wealth that is evenly distributed.[11] In other words, they are the Scandinavian countries: Norway, Sweden, and Denmark, which took the three top places in Delhey and Newton's survey of generalized trust in sixty countries. The researchers find that other countries are trustful to the extent that they share these traits.

Wealth and economic equality promote trust by reducing conflicts of interest. In an affluent society, people will not be drawn into conflict over scarce resources, and if its resources are evenly distributed, the potential for conflict is reduced still further. Economic equality is also a form of similarity, like ethnic homogeneity, which favours trust.

The role of Protestantism is more mysterious. It has loomed large in sociological discussion ever since Max Weber discussed its role in the rise of capitalism, and is held to be associated with hard work, thrift, and enterprise; but possessing several admirable qualities does not guarantee that one possesses others. Delhey and Newton speculate that the association may have arisen because Protestantism has promoted equality and has impressed on cultures the importance of consistently trustworthy behaviour. Gerry Mackie, discussing social trust in Europe for a Russell Sage Foundation volume, proposes deeper roots. He suggests that both Protestantism and trust were favoured by the distinctive marriage pattern of north-western Europe, possibly rooted in Germanic tribal customs and certainly pre-dating the Protestant schisms from the Catholic church.[12] Women married late by comparison with other parts of Europe, and couples would set up new homes rather than remaining in the households of the husbands' families. This helped to break down the defensive perimeter of the family, inhibiting the kind of social distrust apparent in southern Italy. It also gave women the opportunity to accumulate social capital of their own before marriage, which might have promoted trust between spouses by reducing the inequality between them.

Delhey and Newton's results echo those reported by Putnam, whose American survey, quoted in Chapter 5, found trust levels at their highest in a South Dakota county where the gamut of diversity runs from Swedes to Norwegians.[13] They also pose the same question. As Delhey and Newton point out, if people's trust of others depends on ethnic similarity, we should ask 'exactly how general generalized social trust is'.

Norwegians and Swedes might reply that their home countries have allowed in large numbers of immigrants: people with foreign origins comprise over 16 per cent of the Swedish population, while demographic projections suggest that immigrants could account for more than a quarter of Norway's population by the middle of the century.[14] The EU's Nordic citizens are the most positive in the Union about immigration; but the Scandinavians among them are markedly more concerned than most EU populations about the integration of foreigners.[15]

Time will tell whether the Scandinavian social trust model needs to include ethnic homogeneity; but in the meantime Delhey and Newton's international table of trust offers a striking example of a highly trustful, populous country that is ethnically and religiously diverse.[16] In fourth place after the Scandinavian trio is Canada, jointly with the Netherlands. Canadian trustfulness was illustrated by the US polemicist Michael Moore, who crossed the border into Ontario while making his film *Bowling for Columbine*, and found by unsystematic experiment that Ontarians leave their doors unlocked. They thus avoid the incidental costs of mistrust, which mount up as one makes sure the windows are shut, sets the alarm, locks the door, and checks again that one has locked it, unable to relax after this series of reminders of insecurity.

Canada serves as a reminder that ethnic relations are the products of particular histories, and that findings concerning them in the United States may not apply to its immediate north, let alone to the rest of the world. Robert Putnam, whose findings in the United States suggest a dispiriting tendency for people to withdraw from each other in diverse neighbourhoods, points that out himself. He also depicts this 'hunkering down' as a phenomenon of the short to medium term, drawing optimistic lessons about integration from the longer history of US minorities and institutions. 'It would be unfortunate if a politically correct progressivism were to deny the reality of the challenge to social solidarity posed by diversity,' he concludes. 'It would be equally unfortunate if an ahistorical and ethnocentric conservatism were to deny that addressing that challenge is both feasible and desirable.'[17]

It will be more than unfortunate if societies, of whatever ideological composition, fail to address that challenge. Capitalism has won the global game of how to make a living; its winning strategy is based upon the drive for constantly increased efficiency through mobility. Goods and services must be moved in greater quantity across greater distances; and so must the people who produce them. Countless others will also be driven by desperation from places stricken by war or devoid of resources to regions where there is little demand for their services. Everywhere people are set in motion by the dynamism of capital. All regions that are economically successful, or strive to be, will draw migrants to them. Migrants are inevitable, though they may be managed in very different ways. A country may choose to treat incoming labour as it treats incoming goods, as assets for its convenience. Migrant workers can be excluded from society by a range of instruments including, at the extremes, labour camps and domestic slavery. A high degree of exclusion is not tolerable for the excluded, and it should not be tolerable for any decent national conscience. It is not even an especially agreeable condition for the excluders. Having preferred domination to trust, they must maintain a permanent guard in case their menials steal, defect, or rise against them.

Alternatively, societies can work to include their minorities, new and old. Putnam talks of creating shared identities, which encourage the sense that interests also are shared. States have a role to play in citizen-building, as do other institutions; so too do civil societies and citizens themselves. Many if not most of America's voluntary immigrants are eager to become Americans and are proud when they feel they have done so, while proudly incorporating their old labels in their new designations as 'hyphenated Americans'. At least some young Eastern Europeans are keen to become Europeans, rather than to stay wrapped up in their national overcoats as they venture into Europe's labour markets. Different responses will arise in different contexts; but a necessary condition everywhere is that the people concerned must want to make a diverse society work.

Those who do may also need a resolve greater than that of the newspapers, populist parties, and sections of public opinion who dislike the presence of others and adduce any associated problem as evidence that diversity is undesirable. They also need to stand up to the mortal threat from small factions committed to wage war on diversity. On July 7, 2005, several young Yorkshiremen of Pakistani heritage and Islamist ideology exploded bombs on London underground trains and a bus, in an attack that killed over fifty people: the dead were a snapshot of the capital's diversity, their personal or family origins spanning more than a dozen countries across Europe, Africa, Asia, and beyond. Londoners experienced the awful distrust worked by native terror. Whatever their previous disposition towards diversity, they looked at young men of South Asian appearance on Tube trains, and wondered. Suicide bombings are, among other things, attacks on trust.

'All fixed, fast-frozen relations, with their train of ancient and venerable prejudices and opinions, are swept away, all new-formed ones become antiquated before they can ossify.'[18] In this at least Karl Marx and Friedrich Engels were right, and in this at least their depiction of the compulsive transformative energy of capitalism seems truer today than when they wrote the Communist Manifesto in 1848. They saw that it was this all-enveloping drive to change, the '[c]onstant revolutionizing of production, uninterrupted disturbance of all social conditions, everlasting uncertainty and agitation', that defined the modern epoch.

Their fast-frozen relations, backed by venerable beliefs, correspond to the relationships based on traditional authority that Anthony Giddens contrasts with modern relationships, which are fluid, readily formed, and as readily abandoned. When venerable beliefs lose their grip on relationships, trust may take over their role; but lasts only until the relations are swept away. The industrial revolution drew country folk into villages which turned into cities; the folk turned into the working class, which at times achieved levels of solidarity to compare with the proudest of nations. Trust grew from kinship, comradeship, community, and shared interests; it lasted until it was dispelled by a more recent phase of capitalism's constant revolution.

Life in the streets and factories of the old industrial world favoured the growth of thick trust, the kind that grows from, and with, personal knowledge. People grew up in the houses in which they were born; they worked with the same colleagues until they retired or died. The horizons were as narrow as the streets. It was a paradoxical and, taking the longer view, uncharacteristic effect of modern transformation. Thin trust, which relies on reputation and assessments based on signals, is the kind that is in tune with the spirit of the age. Together with the ferment of massively indirect reciprocity that circulates free software, knowledge, and advice around the world, it is what makes the internet go round, creating networks that shrug off ethnicity, gender, and other demographic constraints. It becomes ever more valuable as technological advances and economic imperatives create ever more opportunities to interact with others, in person or the other side of the world. People need to be open to new relationships, especially since old ones are so often swept away.

Under these conditions trust has little time to develop organically, which is why many companies attempt to hothouse it. The faster the staff turnover, and the more frequent the shifts of power caused by internal promotions or reorganization, the greater the need for artificial and slightly hysterical team-building exercises. These efforts are often accompanied by an aversion to risk that would, if displayed by a person rather than a company, also be described as neurotic. Seeking to control risks, employers codify what they once might have trusted their employees to decide for themselves.

At least employees are aware of their relationship with a corporate entity, and of the distinct interests on either side. Outside work people are extensively indifferent, or oblivious, to the implications of relationships involving the transmission of personal information. Countless thousands of people, mostly young, have signed up to social networking websites. Some have learned the hard way that though they can choose their 'friends', they cannot choose their friends' friends, and once they have entrusted indiscretions to others, they cannot delete them. In the real world, this would easily be recognized as a recipe for embarrassment or worse. Nor (until Facebook started circulating details of users' online purchases) has there seemed to be much sense that the providers of these services have interests of their own, which may not entirely coincide with those of subscribers.[19] They seem to be seen as neutral expert systems rather than enterprises which determine the structure of the relationships they enable. This may owe something to the way in which consumer marketing trades on narcissism. Subscribers look at their social networking profiles, and all they see is themselves.

Although telecommunications—meaning communication at a distance, whether by phone, computer, or other means yet to be invented—are technologies attuned to thin trust, it would be remiss to overlook their role in maintaining thick trust in a mobile world. Young migrants can stay in routine touch with their parents back home more easily than their parents would have been able to contact home from a neighbouring town. But though that may have to do as a substitute for a daughter's or a son's presence, it will not work for the grandchildren who are born abroad. Telecommunications can only mitigate the effects of migration on family relationships, in which much of the potential for rich trust resides.

The class of relationships that stands out as conspicuously problematic under prevailing conditions are those involving couples. When these were maintained by a sense of obligation—to children, to the wider family, or to God—partners were forced into a commitment that could, given effort and compromise, build trust between them. Now that relationships are primarily conditional, partners are readier to factor in the opportunity costs. Instead of feeling that they need to invest more heavily in the relationship, they may be more concerned about the opportunities for other relationships they are missing by sticking with the current one. Here thin trust, which allows people to be open to new relationships, biases them against thick trust.

Much of this shift in strategy derives from the liberation of women, who see far more opportunities open to them than their mothers or grandmothers did, and are far less willing to subordinate their interests to those of others, principally their male partners. This is one important respect in which reducing inequality does not promote trust.

It is not typical, however. By and large, developments that reduce inequality favour trust. They also favour health and happiness. Trust is desirable in itself. When it is placed well, it enhances relations of all kinds. Life is more enjoyable, work is more productive, relationships are more meaningful and rewarding. And it is also part of a complex of factors—association, social capital, community, democracy, equality, health, and happiness—that make for a good society. Trust is to be sought for its own sake, and because it keeps good company.

NOTES

Chapter 1: Just Going Round to the Shop

1. Baier, 1986.
2. Hardin, 2004a, ch. 1.
3. Ibid.
4. Ibid.
5. Ashworth, 2000, 30.
6. Connes, 2004, 93–4.
7. Rousseau *et al.*, 1998.
8. Gambetta and Hamill, 2005.
9. 'Forcing Republicans to deliver', available at http://www.dup.org.uk/Forcing.asp

Chapter 2: Trust from the Barrel of a Gun

1. Dawkins and Krebs, 1978.
2. Ibid. 282.
3. Ibid. 283.
4. Ibid. 285.
5. Ibid. 309.
6. Krebs and Dawkins, 1984.
7. Maynard Smith and Harper, 2003, *v*.
8. Zahavi and Zahavi, 1997.
9. Graves, 1960, 188–91.
10. Ashworth, 2000, 126.
11. Ibid. 194.
12. Ibid. 80.
13. Ibid. 146.
14. Hobsbawm and Ranger, 1983, 287.
15. Ashworth, 2000, 116.
16. Ibid. 144.
17. Todman, 2005, 84–5.
18. Axelrod, 1985, 182.
19. Ibid. 173.
20. Ostrom and Walker, 2003, 129.

Chapter 3: Reason to Believe

1. Whiten and Byrne, 1997.
2. Knight, 1999.
3. Berg *et al.*, 1995.
4. Brosnan *et al.*, 2005.
5. Henrich, 2000.
6. Lecky, 1876, 103.
7. Toshio Yamagishi, in Cook, 2001, ch. 4.
8. Hollis, 1998; Robert Sugden, quoted in O'Hagan and Goodenough, 2000, 5.
9. Hollis, 1998, 14–17.
10. Ibid. 22.
11. Ibid. 23. Other translations of *das versprechen darf* render it as 'entitled' or 'permitted to promise'.
12. Ibid. 31–2.
13. Ibid. 32.
14. Frank, 1988, 16–19.
15. Hollis, 1998, 37–8.
16. Ibid. 64.
17. Hume, 2003.
18. Hollis, 1998, 102.

Chapter 4: In God We Trust

1. John 3: 16, King James Version Bible, http://quod.lib.umich.edu/k/kjv/
2. Matthew 6: 31, 33, ibid.
3. Kohn, 2004, 235.
4. Hamilton, 1996, 325.
5. UK Post Office.
6. *Really Useful Guidebook to Princess Margaret Hospital for Children*, available at

http://pmh.health.wa.gov.au/general/visiting/1290.pdf

7. Giddens, 1991, 90.
8. Ibid. 6.
9. Giddens, 1992, 138–9.
10. Jamieson, 1999.
11. Giddens, 1991, 18.
12. 'Public Expects the Impossible from Science', Ipsos MORI, 2 April 2002, available at http://www.ipsos-mori.com/polls/2002/science.shtml
13. 'Immunisation Statistics, England 2005–2006', The Information Centre, available at http://www.ic.nhs.uk/statistics-and-data-collections/health-and-lifestyles/immunisation/immunisation-statistics-england-2005–2006; 'MMR Uptake Rises in London's under 5s', Health Protection Agency, 27 June 2005, available at http://www.hpa.org.uk/london/press/MMR_27jun05.htm
14. ICM, Guardian Poll February 2002, available at http://www.icmresearch.co.uk/pdfs/2002_february_guardian_february_poll.pdf
15. O'Neill, 2002, 142.

Chapter 5: Through Thick and Thin

1. Diego Gambetta, in Gambetta (ed.), 1988, ch. 10, Anthony Pagden, in ibid., ch. 8; Gambetta, 1993.
2. Paolo Mattia Doria, quoted by Pagden in Gambetta (ed.), 1988, ch. 8.
3. Gambetta, in Gambetta (ed.), 1988, ch. 10.
4. Gambetta, 1993, 224.
5. Ibid. 19.
6. Gambetta, in Gambetta (ed.), 1988, ch. 10.
7. Fukuyama, 1999, 241.
8. Fukuyama, 1995, 26.
9. Fukuyama, 1999, 240. In a study discussed in Chapter 7, however, Delhey and Newton, 2005, find that China has the fourth highest generalized trust score of sixty countries surveyed.
10. Gambetta, in Gambetta (ed.), 1988, 160–1.
11. Putnam, 2000, 118.
12. de Tocqueville, 1997, vol. II, ch. 5, available at http://xroads.virginia.edu/~HYPER/DETOC/ch2_05.htm
13. Putnam, 2000, 49.
14. Ibid. 135.
15. Ichiro Kawachi *et al.*, in Kawachi, Kennedy, and Wilkinson (eds), 1999, chs 23, 22.
16. Ibid.; Wilkinson, 2005; Marmot, 2004.
17. Shively and Clarkson, 1994.
18. Sapolsky, 2004.
19. Putnam, 2000, 136.
20. Putnam, 2007.
21. Letki, 2007.
22. The Stephen Lawrence Inquiry, 1999, ch.6, available at http://www.archive.official-documents.co.uk/document/cm42/4262/sli-06.htm#6.6

Chapter 6: The Goodwill of the People

1. Jan Richter, '"Come with Me to Prague"—Kurt Vonnegut and the Czech underground', Radio Prague, available at http://www.radio.cz/en/article/90508

2. 'Trial of Czechoslovak Jazz Section Imminent?', Open Society Archives, available at http://files.osa.ceu.hu/holdings/300/8/3/text/23-5-1.shtml
3. James Wood, 'Bohumil Hrabal', *London Review of Books* 23, 4 January 2001, available at http://www.lrb.co.uk/v23/n01/wood02_.html
4. Tadeusz Mazowiecki, 'Solidarity Takes Power', available at http://www.project-syndicate.org/commentary/maz1
5. Eade *et al.*, 2007.
6. 'Evaluation of Democracy', Polish public opinion, January 2007, CBOS, available at http://www.cbos.pl/PL/Opinia/2007/01_2007.pdf
7. Shin and Wells, 2001; Mishler and Rose, 2005.
8. Mishler and Rose, 2005; Uslaner, 2002.
9. Keele, 2007; Putnam, 2000.
10. Martin Amis, 'The latest bad news', *Observer* 25 October 1987.
11. Secor and O'Loughlin, 2005.
12. Kim, 2005.
13. Freitag, 2006.
14. David Hume, 'Of the Independency of Parliament', in Hume, 1987, available at http://www.econlib.org/library/LFBooks/Hume/hmMPL6.html
15. Hardin, 2002.
16. Jiobu and Curry, 2001.
17. Killias, 1993, gives rates of gun ownership per million as 27.2 in Switzerland and 48 in the United States in 1989, and gun homicide rates per million as 4 in Switzerland and 44.6 in the United States.
18. 'Ban on Army Firearms at Home Wins Support', available at http://www.swissinfo.org, 22 April 2007.
19. 'Brown Promises to "Build Trust"', BBC News, 18 May 2007, available at http://news.bbc.co.uk/1/hi/uk_politics/6667693.stm
20. *2003 Home Office Citizenship Survey: People, Families and Communities*, Home Office Research, Development and Statistics Directorate, 2004, available at http://www.homeoffice.gov.uk/rds/pdfs04/hors289.pdf
21. YouGov/*Daily Telegraph* Survey Results, 28–30 November 2006, available at http://www.yougov.com/archives/pdf/TEL060101024_3.pdf; Peter Kellner, 'ID cards—A "Plastic Poll Tax"?', YouGov.plc: Understanding People, 5 December 2006, available at http://yougov.com/interactive/kellnerMain.asp?jID=3&aId=4183&sID=6&wID=0&UID=. The survey showed overwhelming support for CCTV cameras in public places—85 per cent of respondents welcomed them in high streets—but only 50 per cent approved of speed cameras. Forty-four per cent felt that the state already held too much information about them; 35 per cent felt the amount held was about right, 6 per cent thought it was not enough, and 15 per cent answered 'don't know'. Two-thirds of respondents did not trust governments to keep

information on the national database confidential; 82 per cent foresaw a danger that individual civil servants would divulge database information improperly.
22. 'Q&A: Child Benefit Records Lost', BBC News, 22 November 2007, available at http://news.bbc.co.uk/1/hi/uk_politics/7103828.stm
23. Margaret Levi, in Braithwaite and Levi (eds), 1998, ch. 4.
24. Ibid.: 7 per cent approved; 79 per cent disapproved.
25. 'Young Voter Strategies, Partisanship: A Lifelong Loyalty That Develops Early', February 2007, available at http://www.youngvoterstrategies.org/index.php?tg=fileman&idx=get&id=1&gr=Y&path=Research&file=Partisanship+is+a+Habit.pdf
26. Opinion polls conducted in 2005 and 2007 showed majorities of 52 and 60 per cent, respectively, thought that members of the Bush administration had deliberately misled the public over the case for war with Iraq: CBS News/*New York Times* Poll, 'The War in Iraq', 4–8 September 2007, available at http://www.cbsnews.com/htdocs/pdf/sep07a-iraq.pdf
27. 'The Chairman's Address "On The Threshold Of Victory"', *Times*, 2 October 1962, 16.

Chapter 7: Leaving the Door Unlocked

1. Glaeser *et al.*, 2000.
2. Berg *et al.*, 1995.
3. Delhey and Newton, 2005.
4. Russell Hardin in Hardin (ed.), 2004, 284–6.
5. Rahn and Transue, 1998.
6. Uslaner, 2002.
7. Ibid. ch. 2.
8. Ibid. ch. 9.
9. Rothstein and Uslaner, 2005.
10. Ridley, 1996, 265.
11. Delhey and Newton, 2005.
12. Gerry Mackie, in Cook (ed.), 2001, ch. 8.
13. Putnam, 2007.
14. 'Sweden and Swedes', available at http://www.sweden.se/upload/sweden_swedes/static/en/PopUp.html; Statistics Norway, 'Projection of the Norwegian, Immigrant Population', available at http://www.ssb.no/english/subjects/02/03/innvfram_en/
15. Eurobarometer, 'European Social Reality Report', 2007, available at http://ec.europa.eu/public_opinion/archives/ebs/ebs_273_en.pdf. Finns, Swedes, and Danes were found to hold the most positive views in the EU about immigration. The integration of foreigners was a main current concern for 17 per cent of Swedes and 28 per cent of Danes, compared to an EU average of 8 per cent. Norway is not a member of the EU.
16. Twenty-eight per cent of Canada's population, totalling an estimated 33,390,141, derives from the British Isles, 23 per cent from France, 15 per cent from other parts of Europe; 2 per cent are Amerindian, 26 per cent are of 'mixed background', and others

make up the remaining 6 per cent. The largest religious group is Catholic (42.6 per cent); Protestants comprise the second largest, with 23.3 per cent: Central Intelligence Agency, The World Factbook, available at https://www.cia.gov/library/publications/the-world-factbook/geos/ca.html#People (updated July 2007).

17. Putnam, 2007.
18. Marx and Engels, 2006.
19. 'Protests Force Facebook to Change', BBC News, 30 November.

FURTHER READING

Baier, Annette, 1991, 'Trust', Tanner Lectures on Human Values, University of Utah Press, Salt Lake City, available at http://www.tannerlectures.utah.edu/documents/lectures/baier92.pdf

Frank, Robert H., 1988, *Passions Within Reason: The Strategic Role of the Emotions*, Norton, New York.

Fukuyama, Francis, 1995, *Trust: The Social Virtues and the Creation of Prosperity*, Hamish Hamilton, London.

Gambetta, Diego (ed.), 1988, *Trust: making and breaking cooperative relations*, Blackwell, Oxford, available at https://www.nuff.ox.ac.uk/users/gambetta/gambetta_trust%20book.pdf

Hardin, Russell, 2004b, *Trust and Trustworthiness*, Russell Sage Foundation, New York, chapter 1 available at http://www.russellsage.org/publications/books/0-87154-341-9/chapter1_pdf

Putnam, Robert D., 2000, *Bowling Alone: The Collapse and Revival of American Community*, Simon and Schuster, New York.

BIBLIOGRAPHY

Ashworth, Tony, 2000, *Trench Warfare 1914–1918: The Live and Let Live System*, Pan, London.

Axelrod, Robert, 1985, *Evolution of Co-operation*, Basic, New York.

Baier, Annette, 1983, 'What do Women Want in a Moral Theory?', *Noûs* 19(1):53–63.

Baier, Annette, 1986, 'Trust and Antitrust', *Ethics* 96(2):231–60.

Baier, Annette, 1991, 'Trust', Tanner Lectures on Human Values, University of Utah Press, Salt Lake City, available at http://www.tannerlectures.utah.edu/lectures/baier92.pdf

Becker, Lawrence C., 1996, 'Trust as Noncognitive Security about Motives', *Ethics* 107:43–61.

Berg, J., Dickaut, J., and McCabe, K., 1995, 'Trust, Reciprocity and Social History', *Games and Economic Behavior* 10:122–42.

Braithwaite, Valerie, and Levi, Margaret, (eds), 1998, *Trust and Governance*, Russell Sage Foundation, New York.

Brosnan, Sarah F., Schiff, Hillary C., and de Waal, Frans B. M., 2005, 'Tolerance for Inequity May Increase with Social Closeness in Chimpanzees', *Proceedings of the Royal Society B* 272:253–8.

Coleman, James S., 1990, *Foundations of Social Theory*, Belknap, Cambridge, Mass.

Connes, Georges, 2004, *A POW's memoir of the First World War: The Other Ordeal*, Berg, Oxford.

Cook, Karen S. (ed.), 2001, *Trust in Society*, Russell Sage Foundation, New York.

Cook, Karen S., Hardin, Russell, and Levi, Margaret, 2005, *Cooperation without Trust?*, Russell Sage Foundation, New York.

Dawkins, Richard and Krebs, John R., 1978, 'Animal Signals: Information or Manipulation?', in *Behavioural Ecology: An Evolutionary Approach* (Krebs, John, and Davies, Nicholas, eds), Blackwell, Oxford, pp. 282–309.

Delhey, Jan, and Newton, Kenneth, 2005, 'Predicting Cross-National Levels of Social Trust: Global Pattern or Nordic Exceptionalism?', *European Sociological Review* 21:311–27.

de Tocqueville, Alex, 1997, *Democracy in America*, 2 vols. ed. and trans. Henry Reeve, available at http://xroads.virginia.edu/~HYPER/DETOC/

Eade, John, Drinkwater, Stephen, and Garapich, Michał P., 'Class and Ethnicity: Polish Migrant Workers in London', Centre for Research on Nationalism, Ethnicity and Multiculturalism, University of Surrey, available at http://www.surrey.ac.uk/Arts/CRONEM/polish/POLISH_FINAL_RESEARCH_REPORT_WEB.pdf

Frank, Robert H., 1988, *Passions within Reason: The Strategic Role of the Emotions*, Norton, New York.

Freitag, Markus, 2006, 'Bowling the State Back in: Political Institutions and the Creation of Social Capital', *European Journal of Political Research* 45:123–52.

Fukuyama, Francis, 1995, *Trust: The Social Virtues and the Creation of Prosperity*, Hamish Hamilton, London.

Fukuyama, Francis, 1999, *The Great Disruption: Human Nature and the Reconstitution of Social Order*, Profile, London.

Gambetta, Diego, 1993, *The Sicilian Mafia: The Business of Private Protection*, Harvard University Press, Cambridge, Mass.

Gambetta, Diego, and Hamill, Heather, 2005, *Streetwise: How Taxi Drivers Establish their Customers' Trustworthiness*, Russell Sage Foundation, New York.

Gambetta, Diego (ed.), 1988, *Trust: Making and Breaking Cooperative Relations*, Blackwell, Oxford, available at https://www.nuff.ox.ac.uk/users/gambetta/gambetta_trust%20book.pdf

Giddens, Anthony, 1991, *Modernity and Self-Identity: Self and Society in the Late Modern Age*, Polity, Cambridge.

Giddens, Anthony, 1992, *The Transformation of Intimacy: Sexuality, Love and Eroticism in Modern Societies*, Polity, Cambridge.

Glaeser, Edward L., Laibson, David I., Scheinkman, José A., and Soutter, Christine L., 2000, 'Measuring Trust', *Quarterly Journal of Economics* 115: 811–46.

Graves, Robert, 1960, *Goodbye to All That*, Penguin, London.

Hamilton, W. D., 1996, *Narrow Roads of Gene Land, vol. 1: Evolution of Social Behaviour*, W. H. Freeman, Oxford.

Hardin, Russell, 2002, 'Liberal Distrust', *European Review* 10(1):73–89.

Hardin, Russell, 2004a, *Distrust*, Russell Sage Foundation, New York.

Hardin, Russell, 2004b, *Trust and Trustworthiness*, Russell Sage Foundation, New York.

Henrich, J., 2000, 'Does Culture Matter in Economic Behavior? Ultimatum Game Bargaining among the Machiguenga of the Peruvian Amazon', *American Economic Review* 90(4):973–9.

Hobsbawm, Eric J., and Ranger, Terence O., 1983, *The Invention of Tradition*, Cambridge University Press, Cambridge.

Hollis, Martin, 1998, *Trust within Reason*, Cambridge University Press, Cambridge.

Hume, David, 1987, *Essays, Moral, Political, and Literary*, Liberty Fund, Indianapolis, available at http://www.econlib.org/library/LFBooks/Hume/hmMPL.html

Hume, David, 2003, *An Enquiry Concerning the Principles of Morals*, Project Gutenberg, available at http://www.gutenberg.org/dirs/etext03/nqpmr10.txt

Jamieson, Lynn, 1999, 'Intimacy Transformed? A Critical Look at the "Pure Relationship"', *Sociology* 33:477–94.

Jiobu, Robert M., and Curry, Timothy J., 2001, 'Lack of Confidence in the Federal Government and the Ownership of Firearms', *Social Science Quarterly* 82(1):77–88.

Kawachi, Ichiro, Kennedy, Bruce P., and Wilkinson, Richard G., 1999, *The Society and Population Health Reader, vol. 1: Income Inequality and Health*, New Press, New York.

Keele, Luke, 2007, 'Social Capital and the Dynamics of Trust in Government', *American Journal of Political Science* 51:241–54.

Killias, Martin, 1993, 'Gun Ownership, Suicide and Homicide: An International Perspective', in *Understanding Crime Experiences of Crime and Crime Control* (del Frate, Anna Alvazzi, Zvekic, Ugljesa and van Dijk, Jan J. M., eds), UNICRI, Rome, pp. 289–306, available at http://www.unicri.it/wwk/publications/books/series/understanding/19_GUN_OWNERSHIP.pdf

Kim, Ji-Young, 2005, ' "Bowling Together" Isn't a Cure-All: The Relationship between Social Capital and Political Trust in South Korea', *International Political Science Review* 26:193–213.

Knight, Chris, 1999, 'Sex and Language as Pretend-Play', in: *The Evolution of Culture: An Interdisciplinary View* (Dunbar, R., Knight, C., and Power, C., eds), Edinburgh University Press, Edinburgh, pp. 228–47.

Kohn, Marek, 2004, *A Reason for Everything: Natural Selection and the English Imagination*, Faber, London.

Krebs, John and Dawkins, Richard, 1984, 'Animal Signals: Mind-Reading and Manipulation', in *Behavioural Ecology: An Evolutionary Approach* (Krebs, John, and Davies, Nicholas, eds), Blackwell, Oxford, pp. 380–402.

Lecky, William E. H., 1876, *History of European Morals from Augustus to Charlemagne*, Appleton, New York.

Letki, Natalia, 2007, 'Does Diversity Erode Social Cohesion? Social Capital and Race in British Neighbourhoods', *Political Studies*, OnlineEarly

Articles, doi:10.1111/j.1467-9248.2007.00692.x, available at http://letki.civitas.edu.pl/NLetki_social%20capital%20and%20diversity_final.pdf

Markova, Ivana (ed.), 2004, 'Trust and Democratic Transition in Post-Communist Europe', *Proceedings of the British Academy* 123, Oxford University Press.

Marmot, Michael, 2004, *Status Syndrome: How Your Social Standing Directly Affects Your Health and Life Expectancy*, Bloomsbury, London.

Marx, Karl and Engels, Friedrich, 2006, *Manifesto of the Communist Party*, available at http://www.anu.ed.au/polsci/marx/classics/manifesto.html

Maynard Smith, John, and Harper, David, 2003, *Animal Signals*, Oxford University Press, Oxford.

Milgrom, Paul R., 1984, 'Axelrod's *The Evolution of Co-operation*', *Rand Journal of Economics* 15:305–9.

Mishler, William, and Rose, Richard, 2005, 'What Are the Political Consequences of Trust?: A Test of Cultural and Institutional Theories in Russia', *Comparative Political Studies* 38:1050–78.

Misztal, Barbara, 1996, *Trust in Modern Societies: The Search for the Bases of Social Order*, Polity, Cambridge.

O'Hagan, Timothy, and Goodenough, Jerry (eds), 2000, 'Essays in Honour of Martin Hollis, Arising from a Reading of *Trust within Reason*', UEA Papers in Philosophy New Series, University of East Anglia, Norwich.

O'Neill, Onora, 2002, *Autonomy and Trust in Bioethics*, Cambridge University Press, Cambridge.

Ostrom, Elinor, and Walker, James, (eds) 2003, *Trust and Reciprocity: Interdisciplinary Lessons for Experimental Research*, Russell Sage Foundation, New York.

Putnam, Robert D., 2000, *Bowling Alone: The Collapse and Revival of American Community*, Simon and Schuster, New York.

Putnam, Robert D., 2007, '*E pluribus unum*: Diversity and Community in the Twenty-First Century', The 2006 Johan Skytte Prize Lecture, *Scandinavian*

Political Studies 30:137–74, available at http://www.blackwell-synergy.com/doi/abs/10.1111/j.1467-9477.2007.00176.x

Rahn, Wendy M., and Transue, John E., 1998, 'Social Trust and Value Change: The Decline of Social Capital in American Youth, 1976–1995', *Political Psychology* 19:545–65.

Ridley, Matt, 1996, *The Origins of Virtue*, Viking, London.

Rothstein, Bo, and Uslaner, Eric, 2005, 'All for All: Equality, Corruption, and Social Trust', *World Politics* 58:41–72.

Rousseau, Denise M., Sitkin, Sim B., Burt, Ronald S., and Camerer, Colin, 1998, 'Not So Different After All: A Cross-discipline View of Trust', *Academy of Management Review* 23:393–405.

Sapolsky, Robert M., 2004, 'Social Status and Health in Humans and Other Animals', *Annual Review of Anthropology* 33:393–418.

Secor, Anna J., and O'Loughlin, John, 2005, 'Social and Political Trust in Istanbul and Moscow: A Comparative Analysis of Individual and Neighbourhood Effects', *Transactions of the Institute of British Geographers (New Series)* 30:66–82.

Seligman, Adam B., 2000, *The Problem of Trust*, Princeton University Press, Princeton, NJ.

Shin, Doh C., and Wells, Jason, 2001, 'Testing the Churchill Notion of Democracy as a Lesser Evil in Post-Communist Europe', *Central European Political Science Review* 2:6–24.

Shively, Carol A., and Clarkson, Thomas B., 1994, 'Social Status and Coronary Artery Atherosclerosis in Female Monkeys', *Arteriosclerosis and Thrombosis* 14:721–6.

Skyrms, Brian, 1996, *Evolution of the Social Contract*, Cambridge University Press, Cambridge.

Skyrms, Brian, 2001, *The Stag Hunt*, available at http://www.lps.uci.edu/home/fac-staff/faculty/skyrms/

Sztompka, Piotr, 1999, *Trust: A Sociological Theory*, Cambridge University Press, Cambridge.

Todman, Dan, 2005, *The Great War: Myth and Memory*, Hambledon and London, London.

Uslaner, Eric, 2002, *The Moral Foundations of Trust*, Cambridge University Press, Cambridge, available at http://www.bsos.umd.edu/gvpt/uslaner/research.htm

Whiten, Andrew, and Byrne, Richard W., 1997, *Machiavellian Intelligence II: Extensions and Evaluations*, Cambridge University Press, Cambridge.

Wilkinson, Richard G., 2005, *The Impact of Inequality: How to Make Sick Societies Healthier*, Routledge, London.

Zahavi, Amotz, and Zahavi, Avishag, 1997, *The Handicap Principle: A Missing Piece of Darwin's Puzzle*, Oxford University Press, Oxford.

INDEX